AF573882

Applications of Lithium in Ceramics

Applications of Lithium in Ceramics

J. H. Fishwick

CAHNERS BOOKS, A Division of Cahners Publishing Company, Inc.
89 Franklin Street, Boston, Massachusetts 02110

Library of Congress Cataloging in Publication Data

Fishwick, J H 1932–
Applications of lithium in ceramics.

Includes bibliographical references.
1. Ceramics. 2. Lithium compounds. I. Title.
TP815.F52 666 74–18039
ISBN 0–8436–0611–8

Library of Congress Catalog Card Number: 74–18039
International Standard Book Number: 0–8436–0611–8

Cahners Books, A Division of Cahners Publishing Company, Inc.
89 Franklin Street, Boston, Massachusetts 02110

Printed in the United States of America

Contents

Preface

Recognition of lithia as a powerful flux in ceramic systems, as an important constituent in internally nucleated glass ceramics, and as a low-expansion component in thermal-shock-resistant refractories has resulted in a remarkable growth in the use of lithium chemicals and minerals in the last decade. An increase in the number of papers and abstracts describing the uses of lithium has been evident in the ceramic literature to even the most casual reader.

This book is not intended to be a comprehensive bibliography of all papers published on lithium in ceramics but rather a guideline to those uses of lithium that are felt to be practical in nature. For example, numerous papers have been published on glass compositions containing relatively small amounts of lithia. Undoubtedly lithia plays a significant role in the melting and forming of such glasses and also on the resulting physicochemical properties; however, when such papers have not been included in the bibliography, it is because lithium was only of incidental importance. There have also been many papers written on the crystallization of glasses in systems containing lithia, but I have restricted my examples to those that appear to have some commercial value or to those that illustrate the inherent value of the lithium ion. Finally, I have not included a large number of phase diagrams depicting binary and ternary systems containing lithium. This is intentional. Phase diagrams of lithia systems have been well summarized by Levin, Robbins, and McMurdie (American Ceramic Society) and need no further clarification here.

Applications of Lithium in Ceramics

1
Introduction

Lithium is the lightest metal known. Its atomic number is 3 and its atomic weight is 6.939. It was discovered in 1818 by Arfvedson and named by Berzelius.

In many respects, lithium behaves more like an alkaline earth element than an alkali. The similarity between the atomic and ionic radius of the lithium and magnesium ion is shown below:

	Lithium	*Magnesium*
Atomic radius	1.33 Å	1.36 Å
Ionic radius	0.60 Å	0.65 Å

This similarity is reflected in the properties exhibited by many ceramic systems. For example, the beta spodumene solid solution series has no counterpart in the other alkali systems but is similar to the solid solution series in the MgO-Al_2O_3-SiO_2 system. Like alkaline earths, lithia raises the surface tension of glasses and enamels, while sodium and potassium reduce it. Other similarities between lithium and the alkaline earth elements can be found in the field of organic chemistry, particularly in the organometallic compounds.

Development of Lithium

A curve showing the lithium production in the United States, as pounds of LiOH equivalents, is shown in Figure 1.1.[1] As Shay has pointed out, the lithium industry in the United States has followed a pattern of several periods of rapid growth that were succeeded by problems of overcapacity and lower prices.[2] These factors stimulated the development of additional uses, resulting in another expansion. The first production of lithium on any appreciable scale was just before 1900 with the development of the spodumene Etta Mine in the Black Hills of South Dakota. The first peak in Figure 1.1 (1) reflects the mining of lepidolite bodies at the Stewart Mine in San Diego County, California. Production remained virtually unchanged for several years. Then,

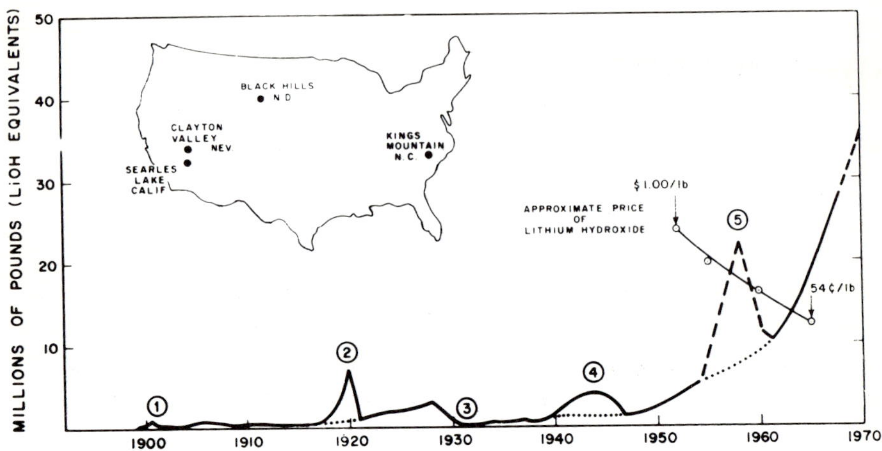

Figure 1.1. Lithium Production in the United States (as pounds of LiOH Equivalents)

renewed activity at the Stewart Mine, combined with the first use of lithium in alkaline storage batteries during World War I, resulted in the second peak (2) shown in Figure 1.1. Following the economic depression (the third peak), military requirements for lithium products during World War II, for such applications as carbon dioxide absorption in submarines and gas masks, resulted in the increased production reflected by peak (4) in Figure 1.1. The rapid growth in lithium production since World War II is a result of the discovery of lithium greases and the development of the air conditioner market. The large increase in lithium chemical production in the middle and late 1950s (peak 5) was due to the requirements of the Atomic Energy Commission for its thermonuclear program. After completion of the last AEC contract in 1960, United States producers were left with excess capacity and reduced profit margins. The subsequent lithium slump caused Maywood Chemical Company to withdraw from the lithium business in the early 1960s and resulted in the closing down of chemical grade spodumene production by the Quebec Lithium Corporation in 1965.

Since the AEC contracts expired in 1960, new applications for lithium in ceramics, multipurpose greases, synthetic rubber, and aluminum potlines have imparted new impetus to the growth curve as indicated in Figure 1.1. This strong growth rate is expected to continue throughout the 1970s, with ceramics being the largest single market for lithium concentrates and compounds.

Uses of Lithium

Apart from its uses in ceramics, with which this book is primarily concerned, lithium is used in many diverse applications from synthetic rubber to high-energy electric batteries. Some of these uses are described briefly below.

Lubricants. Lithium stearate is the principal metallic soap component of multipurpose greases. These greases retain their lubricating properties over a wide temperature range and have good water resistance. About a pound of lithium hydroxide monohydrate is used in forty-five to a hundred pounds of grease.

Synthetic Rubber. Butyllithium is extensively used as a stereospecific catalyst in the production of synthetic rubber. It is also used as a metalating agent in the synthesis of aromatic organolithium compounds.

Air Conditioning. Lithium bromide is used in the form of a concentrated brine for absorption refrigeration machines. The advantages of using a hydroscopic salt such as lithium bromide are:

1. The high latent heat of vaporization of water, which is the refrigerant used in this application, is utilized.
2. Lithium bromide is nonvolatile.
3. The refrigerating system operates at low pressure.
4. The materials are relatively nontoxic.

Industrial Drying. Lithium chloride is one of the most hygroscopic inorganic compounds known, and a saturated solution will dry air in contact with it down to an extremely low humidity. This phenomenon is used extensively in industrial drying systems.

Carbon Dioxide Absorbent. Lithium hydroxide granules are being used as a carbon dioxide absorbent in submarines, space vehicles, and portable life support systems. Factors that affect the rate of carbon dioxide absorption have been discussed in detail by Boryta and Maas.[3]

Welding and Brazing. Lithium chloride and lithium fluoride are used as scavengers and cleaners in welding and brazing, where they form low-melting slags with many oxides. They can also reduce the surface tension of molten metal and increase wettability.

Metallurgy. Lithium is used in only minor amounts for metallurgical uses. Small quantities are added to copper alloys and high-conductivity copper as a deoxidizer to restore the electrical conductivity.

Pharmaceutical. Lithium compounds are used in the synthesis of vitamin A and in the treatment of manic-depressive disorders. Lithium carbonate is used in the latter instance in the amount of about one to one and a half grams per day. The use of lithium in psychiatry has been well documented by Maletzky and Blachly.[4]

Batteries. Considerable effort is currently being expended in developing high-energy density batteries for vehicular applications and military power sources. Because of its high energy density, lithium is a desirable electrode material. Compared with lead, lithium has great theoretical advantages for batteries suitable for an electric automobile. While lithium weighs only about a thirtieth as much as an equal volume of lead, it can generate up to eight times as much electricity.

Argonne National Laboratory is developing a lithium-sulfur battery for automobile propulsion.[5] Reports indicate that it can deliver more than a hundred watt-hours per pound—or five times more than that of lead-acid batteries. The Argonne unit has a lithium anode and a sulfur cathode, while the electrolyte is an eutectic mixture of lithium chloride, potassium chloride, and lithium fluoride.

Aluminum Potlines. Lithium carbonate or lithium fluoride are used in relatively large quantities for increasing the yield of aluminum potlines. The lithium compound (usually lithium carbonate because it is cheaper) is added to the cells where it reacts with the cryolite. This results in lower fuel consumption and a higher current efficiency.

Distribution of Lithium by End Use. Roy and Chakravorty gave the following approximate distribution of lithium chemicals in 1969: [6]

Ceramics and glass	61%
Lubricants and greases	20%
Air conditioning	7%
Pharmaceuticals/chemicals/alloys	12%

Cummings has estimated the demand for lithium in the United States for the year 2000 as a median of 10,900 short tons.[7] The estimate for the rest of the world is 6,900 short tons. Her estimated demand for the United States in the year 2000 by end use is shown in Table 1.1.

It should be noted that these forecasts are of a "surprise-free" nature. The development of new applications for lithium, such as the present use in aluminum potlines, or of a possible future demand for electric batteries, could markedly change the forecast demand.

Table 1.1

	U.S. Demand for Lithium in the Year 2000 (Short Tons)		
End Use	*Low*		*High*
Ceramics and glass	1,000		3,300
Greases	2,700		4,000
Air conditioning	650		1,200
Metallurgy	950		1,900
Welding and brazing	1,750		3,000
Other uses	550		800
Total	7,600		14,200
Adjusted range	8,700		13,100
		(Median 10,900)	

Notes

1. B.J. O'Neill, I.A. Kunasz, and L.A. Wright, "The Lithium Production Curve: A Measure of Many Things," *Earth and Mineral Sciences* 38 (March 1969).

2. F.B. Shay, "Low Cost Lithium From Brine at Silver Peak, Nevada," paper presented at the 96th AIME Annual Meeting in Los Angeles, California, February 21, 1967.

3. D.A. Boryta and A.J. Maas, "Factors Influencing Rate of Carbon Dioxide Reaction with Lithium Hydroxide," *I and EC Process Design and Development* 10 (1971): 489–494.

4. B. Maletzky and P.H. Blachly, *The Use of Lithium in Psychiatry* (Cleveland, Ohio: CRC Press, 1971).

5. E.J. Cairns, P.A. Nelson, E.C. Gay, R.K. Steunenberg, W.L. Walsh, J.E. Battles, H. Shimotake, J.P. Akerman, M.L. Kyle, and D.S. Webster, "Development of High-Specific-Energy Batteries for Electric Vehicles," Report ANL–7998 (Argonne National Laboratory, 1973). M.L. Kyle, E.J. Cairns, and D.S. Webster, "Lithium/Sulfur Batteries for Off-Peak Energy Storage, A Preliminary Comparison of Energy Storage and Peak Power Generation Systems," Report ANL–7958 (Argonne National Laboratory, 1973).

6. S.B. Roy and S.K. Chakravorty, "Lithium in Ceramics–A Review," *Transactions of the Indian Ceramic Society* 29 (1970): 134–144.

7. A.M. Cummings, *Lithium,* Bureau of Mines Bulletin 650, Mineral Facts and Problems (Washington, D.C.: U.S. Department of the Interior, 1970).

2 Raw Materials

Lithium minerals occur predominantly as pegmatites, which are coarse-grained igneous rocks formed by the crystallization of postmagmatic fluids. Mineralogically, granitic pegmatites consist of feldspar, quartz, and mica as the principal minerals and a variety of lithium, beryllium, tantalum, tin, and cesium minerals as the minor constituents. Although lithium occurs in about 145 minerals, the principal commercial sources are spodumene, lepidolite, petalite, and amblygonite. The discovery of Searles Lake in 1938 gave rise to a completely new source for lithium chemicals.

Geochemistry

The geochemistry of lithium was studied and summarized by Goldschmidt, Rankama and Sahama, and Horstman.[1] More recently, Kunasz has discussed the geochemistry of lithium in his report on lithium raw materials. According to Kunasz, the presence of the lithium ion in the crystal lattice of minerals is determined by its size and charge.[2] Its distribution in igneous rocks is controlled by the ratio $(MgO + FeO)/Li_2O$. This ratio is very large in the early stages of crystallization of a magma and results in the removal of ferromagnesium minerals in preference to lithium. The lithium is then concentrated in the residual magma, which results in its enrichment in silicic rocks and pegmatites.

Lithium is also found in small amounts in many different rock types. For example, the lithium content of igneous rocks is about 0.006% Li_2O, whereas sedimentary rocks contain an average of 0.0115% Li_2O, the highest value being recorded in shales. Kunasz also points out that the clay mineral hectorite contains about 1.14% Li_2O and that other lithium-bearing clays occur near Amboy (0.50% Li_2O) and near Boron (0.41% Li_2O) in California, at Spor Mountain, Utah (0.43% Li_2O), in Yavapai County, Arizona (0.21% Li_2O), and in Clayton Valley, Nevada (0.70% Li_2O).[3]

Lithium also occurs in relatively large amounts in spring waters associated with geothermal areas, in oil well brines, and certain brines of California (Searles

Lake), Nevada (Clayton Valley), Utah (Great Salt Lake), and Chile (Salar de Atacama), where it has been concentrated by solar evaporation.[4]

Lithium Pegmatites

According to Norton, lithium pegmatites can be divided into two categories: deposits of spodumene that are virtually homogeneous, and deposits of spodumene or other lithium minerals that are segregated into a series of zones of different composition or texture, in places accompanied by replacement bodies and fracture-filling units.[5]

Foote Mineral Company's deposit at Kings Mountain, North Carolina, is a good example of a homogeneous lithium-rich pegmatite. This deposit has been described in detail by Kesler. The lithium pegmatites contain five minerals in more than trace quantities. The mineral content is given by Kesler as: [6]

Mineral	*Weight Percent*
Spodumene	20
Quartz	32
Muscovite	6
Feldspars (potash and soda)	41
Trace minerals (by difference)	1
	100

The distribution of lithia in the Kings Mountain pegmatite minerals is shown in Table 2.1.

Table 2.1
Distribution of Lithia in the Kings Mountain Pegmatite Minerals

Mineral	*Average* Li_2O *in Mineral (Weight %)*	*Average* Li_2O *in Pegmatite (Weight %)*	*Percent by Weight of Total* Li_2O
Spodumene	7.46	1.492	97.6
Microcline	0.12	0.017	1.1
Albite	0.01	0.003	0.2
Quartz	0.02	0.006	0.4
Muscovite	0.19	0.011	0.7
Trace minerals	—	—	—
Totals		1.529	100.0

The lithium pegmatites occur along parts of the east side of an irregular batholith known as the Cherryville quartz monzonite of probable Devonian age. The pegmatites are enclosed in metamorphic rocks of either thin-layered amphibolite, which consists of hornblende and andesine, or mica schist, which consists mostly of muscovite. The deposit consists of many irregularly shaped fingers of varying size. Crystals of spodumene and microcline are rarely three feet in length, and most are less than a foot long. Spodumene is the principal mineral marketed and is available in several grades. Other ore values that are marketed are feldspar, mica, and a mixture of feldspar and quartz. In addition to these major minerals, the pegmatites contain such unusual minerals as switzerite, eakerite, and lithiophosphate.

Kunasz cites the total ore reserve of the Kings Mountain open-pit mine as 36 million tons, of which nearly 21 million tons are measured and about 15 million tons are indicated.[7]

Lithium Corporation of America, a subsidiary of Gulf Resources Corporation, has developed an open-pit mine near Cherryville, North Carolina. As a result of drilling, the measured reserves have been computed at 14 million tons.

Lithium Brines

Considerable amounts of lithium are being obtained from brines located in the arid deserts of the western United States. Brines are beneficiated at Searles Lake, California, and in Clayton Valley, Nevada.

Lithium has been obtained since 1938 from Searles Lake. Here, the brines occur in two salt beds, an upper and a lower one. The upper beds assay about 0.015% Li_2O. The reserves vary according to the thickness used in the computation. Norton and Schlegel estimate the reserves at 9 million units (20 pounds per unit), while Kesler indicates 7.5 million units.[8] The lower salt bed is thinner than the upper bed and averages 0.006% Li_2O. The reserves of this bed are not known.

In Clayton Valley, Nevada, lithium-bearing brines occur in an undrained structural depression filled with Quaternary sediments, consisting mainly of clay minerals. The brine occurs as a concentrated sodium chloride solution containing subordinate amounts of potassium and minor amounts of magnesium and calcium. The average composite concentration of lithia pumped into the ponds is 0.065% Li_2O. Shay has described the Foote Mineral Company brines at Silver Peak in the Clayton Valley, and he reported the saline brine analysis shown in Table 2.2.[9] As Shay points out, the Mg/Li ratio in this brine is about unity, whereas in most other brines the magnesium content is relatively high. The low

Foote Mineral Company's spodumene mine at Kings Mountain, North Carolina

Aerial view of Foote Mineral Company's spodumene mine at Kings Mountain, North Carolina

Foote Mineral Company's brine ponds at Silver Peak, Nevada

Foote Mineral Company's brine ponds at Silver Peak, Nevada

Foote Mineral Company's low-iron spodumene plant at Kings Mountain, North Carolina

Table 2.2
Saline Brine Analyses

	Ocean %	*Dead Sea %*	*Great Salt Lake %*	*Bonneville %*	*Salton Sea %*	*Silver Peak %*
Na	1.05	3.0	7.0	9.4	5.71	6.2
K	0.038	0.6	0.4	0.6	1.42	0.8
Mg	0.123	4.0	0.8	0.4	0.028	0.04
Li	0.0001	0.002	0.006	0.007	0.022	0.04
Ca	0.040	0.3	0.03	0.12	2.62	0.05
SO_4	0.25	0.05	1.5	0.5	0.00	0.71
Cl	1.900	16.0	14.0	16.0	15.06	10.06
Br	0.0065	0.4	0.0	0.0	0.0	0.0
Li/Mg	1/12720	1/2000	1/135	1/60	1/1.3	1/1
Li/K	1/3800	1/300	1/70	1/90	1/71	1/20

magnesium content permits extraction of lithium without a magnesium operation.

Barrett and O'Neill have described the recovery of lithium from the saline brines of Silver Peak.[10] The brines are pumped from beneath a playa surface inside a closed basin. The playa consists of mixtures of clays, silts, and sands, saturated with saline brines down to the bedrock formation, which is 500 to

1,500 feet below the surface of the playa. The origin of the Silver Peak deposit is evidently related to volcanic activity, and the area is characterized by hot springs, cinder cones, and lava deposits. The brine pumped from wells contains about 300 ppm of lithium and 10–15 weight percent of other dissolved solids. The brines are pumped into a series of solar evaporation ponds, and, after they reach saturation, salts are precipitated in the series: NaCl; a mixture of NaCl and glaserite ($K_3Na(SO_4)_2$); and finally NaCl, glaserite, and KCl. The evaporation rate in this region of Nevada is above fifty inches per year, while the rainfall is usually three inches or less. These conditions concentrate the brine in the ponds to approximately 5,000 ppm. Lithium is then recovered from the brine by precipitation as lithium carbonate.

The composition of a typical well brine is compared to that of the final rich brine in Table 2.3. The lithium concentration of the final rich brine is

Table 2.3
Silver Peak Brines

Element	*Well Brine Weight %*	*Rich Brine Weight %*
Na	6.2	7.8
K	0.8	4.8
Mg	0.04	0.007
Li	0.04	0.50
Ca	0.05	0.004
SO_4	0.71	2.9
Cl	10.1	16.1

Source: W.T. Barrett and B.J. O'Neill, Jr., "Recovery of Lithium from Saline Brines Using Solar Evaporation," in J.L. Rau and L.F. Dellwig, eds., *Third Salt Symposium* (Cleveland: The Northern Ohio Geological Society, Inc., 1970).

5,000 ppm, which represents a twelvefold increase over the well brine concentration and an eightfold increase over the point at which the brine initially becomes saturated.

Foote Mineral Company is presently operating about fifty wells ranging from 500 to 600 feet in depth. Kunasz estimates the in-place reserves to a depth of 1,000 feet at 117 million tons of Li_2O.[11]

Major Lithium Minerals

The minerals representing the major commercial sources of lithium are spodumene, petalite, lepidolite, and amblygonite. The approximate chemical composition and lithia content of these minerals is shown in Table 2.4.

Table 2.4
Composition of Commercial Lithium Minerals

Mineral	*General Formula*	Li_2O *Content (%)*
Spodumene	$Li_2O.Al_2O_3.4SiO_2$	4–7
Petalite	$Li_2O.Al_2O_3.8SiO_2$	3.5–4
Lepidolite	$K_2(Li,Al)_{5-6}[Si_{6-7}Al_{2-1}O_{20}](OH,F)_4$	3–4
Amblygonite	$LiAl(PO_4)(F,OH)$	8–9

Spodumene. Spodumene is a lithium aluminosilicate having the general formula $Li_2O.Al_2O_3.4SiO_2$. It is a monoclinic pyroxene and exhibits pronounced cleavage along the (110) plane, resulting in the production of lath-shaped particles upon fracture. Spodumene undergoes an irreversible phase transformation at about 1000°C. This transformation is accompanied by a 30% volume increase as the specific gravity changes from 3.1–3.2 to about 2.4. The resulting beta spodumene is tetragonal and exhibits a remarkably low thermal expansion coefficient. The color of spodumene is variable; it is nearly white in low-iron varieties and dark green in iron-rich crystals.

Hiddenite ranges in color from clear yellow to emerald green and is considered a gemstone. Other gemstones are triphane, a yellow variety, and kunzite, a transparent lilac-pink variety.

Since the United Nations sanctions were imposed on Rhodesia in 1967, spodumene has become the most important source of lithium in the United States. Spodumene is mined by Foote Mineral Company and Lithium Corporation of America from the lithia pegmatites of North Carolina. Lithium Corporation processes the spodumene concentrates for their lithia values, while Foote Mineral Company sells spodumene concentrates as well as the processed lithium chemicals.

Although the theoretical lithium content of spodumene is 8.03% Li_2O, the spodumene concentrates sold by Foote Mineral Company vary from about 6% to 7% Li_2O depending on purity. The chemical composition of these concentrates is shown in Table 2.5.

Petalite. Petalite is a lithium aluminosilicate; its general formula is $Li_2O.Al_2O_3.8SiO_2$. It forms a series of solid solutions with beta spodumene. Like spodumene, petalite is a monoclinic mineral. Its color is grayish-white and, more rarely, pink. It has two cleavage directions, which form an angle of 38.5°. The basal cleavage is perfect. The specific gravity is 2.42.

The theoretical lithia content of petalite is 4.88% Li_2O. In actual com-

Table 2.5
Chemical Composition of Commercial Spodumene Concentrates

Oxide	*Chemical Grade Spodumene*	*Ceramic Grade Spodumene*	*Low-Iron Spodumene*
Li_2O	6.00	6.88	6.83
Fe_2O_3	2.2	0.90	0.10
Na_2O	0.80	0.32	0.35
K_2O	1.00	0.60	0.14
Al_2O_3	25.70	26.15	26.25
SiO_2	63.50	63.70	64.80

Table 2.6
Typical Analysis of Rhodesian Petalite

Oxide	*Weight Percent*
Li_2O	4.3
K_2O	0.39
Na_2O	0.16
Fe_2O_3	0.06
Al_2O_3	16.80
SiO_2	77.0
LOI	0.80

mercial deposits, the lithia concentration varies from 3.0% to 4.7% Li_2O. A typical analysis of Rhodesian petalite is given in Table 2.6.

Lepidolite. Lepidolite is a phyllosilicate with the general formula $K_2(Li,Al)_{5-6}[Si_{6-7}Al_{2-1}O_{20}](OH,F)_4$. Like spodumene and petalite, lepidolite is monoclinic. Its crystals are short prismatic but very rare. It has perfect basal cleavage. The specific gravity is 2.8 to 2.9.

Lepidolite usually occurs as massive aggregates of small flakes, although it is sometimes found in large tabular crystals. Lepidolite is commonly purple, but rose and bluish varieties are known. The lithia concentration is typically 3–4% Li_2O.

Amblygonite. Amblygonite is a pegmatite mineral with the general formula $LiAl(PO_4)(F,OH)$. It occurs in white to gray masses, usually cleavable or columnar. The specific gravity of amblygonite is 3 to 3.1. Although it may contain as much as 10.2% Li_2O, commercial ores usually contain between 8% and 9% Li_2O.

World Production of Lithium

The production of lithium in the free world is shown in Table 2.7 (from the Minerals Yearbook of the Bureau of Mines). Three countries produce the bulk of lithium raw materials: the United States, Rhodesia, and the Territory of Southwest Africa. Of these, the leading producer is the United States. As Kunasz points out, the Soviet Union is undoubtedly an important producer, although production figures are not available.[12]

United States. There are three main areas in the United States for the production of lithium ores and lithium chemicals. Spodumene is mined in the tin-spodumene belt of North Carolina, while brines are evaporated at Searles Lake, California, and at Silver Peak, Nevada.

The tin-spodumene belt of North Carolina represents the largest reserve of lithium in the free world. These spodumene-bearing pegmatites are mined by Foote Mineral Company and Lithium Corporation of America. The Foote open-pit mine is located about one mile south of the town of Kings Mountain. Eight pegmatites have been recognized within the mine, two of which are well defined and constitute the eastern (peg 1) and western (peg 8) limits of the ore body. The remaining six pegmatites constitute the middle zone.

The mine belonging to Lithium Corporation of America is located near Cherryville, North Carolina. This spodumene-containing pegmatite dips about 15° east and is reported to be 250–300 feet thick.

The original source of lithium in the United States was the Etta Mine in the Black Hills of South Dakota, where spodumene was produced almost continuously for fifty years. Today, only small amounts of amblygonite and spodumene are being produced as by-products of quartz and feldspar. According to Kunasz, several other districts in the United States have produced lithium ores in the past.[13] These include the Pala district of southwestern California (lepidolite, amblygonite), the White Picachio district in Arizona (spodumene, amblygonite), the Harding Mine in Taos County (spodumene, lepidolite), and the Pedlite Mine in Mera County (lepidolite) of New Mexico, the Quartz Creek district in Gunnison County, Colorado (lepidolite), and several areas in New England.

The lithium brines have been described earlier in this chapter. Foote Mineral Company operates solar evaporation ponds at Silver Peak, Nevada, and Kerr-McGee controls the Searles Lake brines, which were previously beneficiated by American Potash and Chemical Corporation.

Rhodesia. A pegmatite about 5,100 feet in length and 95 to 210 feet in width is mined by Bikita Minerals (Pvt) Ltd. in the Bikita Tinfields about forty-five miles east of Fort Victoria. Kunasz describes the pegmatite as being asymmetrically zoned and containing a variety of commercially important lithium minerals as well as beryl and pollucite.[14] Zones of lepidolite and petalite exist, and certain portions of the pegmatite contain mixed units consisting of eucryptite, petalite, and spodumene.

The proven ore reserve is 5.984 million tons averaging 2.90% Li_2O. Petalite is the principal ore in the Al Hayat sector. The 1,964,274 tons of ore averaging 4.05% Li_2O make it the largest single petalite deposit in the world. This sector also contains 492,046 tons of spodumene averaging 4.24% Li_2O, as well as the world's only commercial source of eucryptite.

Southwest Africa. The Karibib district, located about 120 miles from Walvis Bay, contains several strongly zoned pegmatites containing lepidolite, petalite, and small amounts of amblygonite. Lepidolite reserves suitable as flotation feed material are estimated at a million tons,[15] while petalite reserves consist of about 200,000 tons of crude ore.[16]

Canada. There are only two districts in Canada known to contain commercial quantities of lithium minerals: the Preissac-Lacorne district of western Quebec and the Bernic Lake area of Manitoba. In the former district, the largest reserves occur on the Quebec Lithium Corporation property where reserves are estimated at 15 million tons of ore containing 1.2% Li_2O.[17] This operation is presently inactive. The Bernic Lake property is operated by Chemalloy Minerals Ltd. There, the spodumene reserves have been measured at 5 million tons containing 2.89% Li_2O. In addition to spodumene, three lepidolite zones averaging 1.88% to 2.81% Li_2O provide an additional 107,000 tons of ore.

South America. Small amounts of lithium minerals are produced in Brazil and Argentina. In Brazil, lithium-bearing pegmatites are found in the district of Minas Gerais and in the northeastern part of the country. The pegmatites have been mined for cassiterite, tantalite, and beryl, while lithium minerals have been recovered as by-products. In Argentina, lithium pegmatites occur in the Sierras Pampaneas. Here, the pegmatites are zoned and contain spodumene. The reserves consist of about 18,000 tons as spodumene.[18]

Africa. Lithium minerals are mined in small amounts in the Republic of South Africa, Mozambique, Uganda, and Ruanda. The Otavi Mining Company (Pty) Ltd. in the Transvaal recovers minor amounts of petalite, spodumene, and lepidolite. Lepidolite is mined from zoned pegmatite bodies in the Alto Ligonha area of Mozambique. Large amblygonite masses are found in the pegmatite districts in Ruanda and Uganda.[19]

Australia. The only active lithium mining area in Australia is in the Coolgardie district of western Australia where small quantities of petalite and minor amounts of amblygonite are produced.

Table 2.7
Lithium Minerals: Free World Production (Short Tons)

Country	Mineral Produced	1959	1960	1961	1962	1963	1964	1965	1966	1967	1968	1969	1970
North America													
U.S.		W	W	W	W	W	W	W	W	W	W	W	W
Canada[1]	Spodumene	1,378	102	268	250	322	528	507	127	218		200	467
South America													
Argentina	Lithium Minerals	187	NA	NA	496	1,583	799	686	298	272	140	388	
Brazil[2]	Spodumene	468			165	28		7,512	110	6,800		1,709	4,017[4]
	Amblygonite	590	55					28					
Surinam	Amblygonite			475	827	568	NA	NA	NA	NA	NA	NA	NA
Europe													
Spain	Amblygonite		28	19									
Africa													
Mozambique	Lepidolite	99	1	170	302	115		83	NA	276	824	461	111
Rhodesia, Southern[3]	Amblygonite			86	35	52			NA	NA			
	Eucryptite		1,334	1,879	866	1,164	806	705[e]	NA	NA			
	Lepidolite	57,901[2]	15,485	24,037	21,244	16,157	22,943	17,700[e]	NA	NA	67,000[e]	67,000[e]	67,000[e]
	Petalite		63,336	27,698	21,704	29,746	36,447	29,900[e]	NA	NA			
	Spodumene		7,690	1,627	1,496	2,235	6,965	15,300[e]	NA	NA			
Ruanda	Amblygonite	2,965	2,569	1,854	359	406	325		NA				
S. Africa, Republic of	Spodumene	10	173	260	1,263	417	179	958	337		41	39	10
Southwest Africa	Amblygonite	242	161	136	141	128	13	39	30	NA			
	Lepidolite	2,168	972	1,418	1,781	86	407	298	365	NA	1,340	4,372	7,616
	Petalite	2,787	3,909	2,540	1,008	865	798	1,332	1,344	NA			
Uganda	Amblygonite			26	22	53	22	22	78[2]	49	49	NA	NA

Oceania												
Australia	Petalite	18	141	94	437	233						
	Amblygonite			31	22		347	1,112	747	827	795	820
	Spodumene			26	24	58						

Source: U.S. Bureau of Mines, *Minerals Yearbooks.*

1. Li_2O content.

2. Exports.

3. Data not available since 1964.

4. Petalite and lepidolite–government export data.

e: Estimated.

W: Data withheld.

NA: Data not available.

Extraction of Lithium

Lithium compounds are presently commercially processed from spodumene by either an acid treatment or an alkali treatment. The acid treatment, which is the process used by Lithium Corporation of America, is described in detail in a patent by Ellestad and Leute.[20] The alkali treatment, involving calcination of spodumene with limestone, was adopted by Foote Mineral Company.

The Acid Treatment (Lithium Corporation of America). In this process, natural or alpha-spodumene is converted to the more reactive beta-spodumene by heating in a rotary kiln to 1075–1100°C. The kiln discharge is cooled and ball-milled to minus 100 mesh. This milled powder is mixed with sulfuric acid in a slight excess of lithium equivalents and heated to 200–250°C. In the reaction that follows, hydrogen replaces lithium, forming soluble lithium sulfate according to the following equation:

$$Li_2O.Al_2O_3.4SiO_2 + H_2SO_4 \xrightarrow{200\text{–}250^\circ C} H_2O.Al_2O_3.4SiO_2 + Li_2SO_4.$$

The ore is then leached with water, and the excess acid is neutralized with ground limestone. After filtering, an impure lithium sulfate solution saturated with calcium sulfate is obtained. Calcium and magnesium are removed by treatment with hydrated lime and soda ash. The pH of the purified solution is adjusted to 7–8 with sulfuric acid, and the resulting solution is concentrated in an evaporator. Following filtration, lithium carbonate is precipitated at 90–100°C by the addition of a soda-ash solution.

The Alkali Treatment (Foote Mineral Company). In the alkali process, spodumene concentrate (obtained by the froth flotation of crushed and milled pegmatite) is mixed with ground limestone in the ratio of 1 part concentrate to 3.5 parts limestone and rotary calcined in a coal-fired kiln at about 1037°C. The kiln discharge is ground and hot leached, resulting in the production of an impure solution of lithium hydroxide. This is then subjected to evaporation-crystallization to produce lithium hydroxide.

Other Extraction Processes. While the acid and alkali treatments briefly described above represent the two principal commercial methods for extracting lithium from ores, other processes have been used in the past with varying degrees of success. For example, Quebec Lithium Corporation used to calcine natural or alpha-spodumene at 1075–1100°C to convert it to beta spodumene, which was then autoclaved at 190–235°C with an excess of sodium carbonate solution.[21] The resulting slurry was cooled and treated with carbon dioxide to convert lithium carbonate to soluble lithium bicarbonate, which, after filtration, was decomposed by heat to precipitate lithium carbonate. According to Archambault, the inventor of this process, the lithium minerals that are amenable to treatment are spodumene, petalite, eucryptite, and lepidolite. Other interesting processes Archambault patented include the direct production of lithium borate by leaching a calcined lithium-bearing silicate under pressure with sodium tetraborate solution,[22] the extraction of lithium by hydrothermal reaction with sodium hydroxide, sodium pyroborate, sodium orthoborate, sodium sulfides, and sodium silicates,[23] and the production of lithium salts by extracting calcined minerals with sodium bicarbonate and sodium sesquicarbonate.[24]

Summary

Although lithium is present in a great many minerals, the only commercial source in the United States is spodumene. This is mined by both Foote Mineral Company and Lithium Corporation of America from the tin-spodumene belt of North Carolina. The other main sources of lithium chemicals are the lithium brines of Searles Lake, California, and of Clayton Valley, Nevada. These deposits are exploited by Kerr-McGee and Foote Mineral Company respectively.

Notes

1. V.M. Goldschmidt, "The Principles of Distribution of Chemical Elements in Minerals and Rocks," *Journal of the Chemical Society* (1937): 655–673. K. Rankama and T.G. Sahama, "The Alkali Metals: Lithium, Sodium, Potassium, Rubidium, Cesium," *Geochemistry* (Chicago: University of Chicago Press, 1950): 422–442. E.L. Horstman, "The Distribution of Lithium, Rubidium, and Cesium in Igneous and Sedimentary Rocks," *Geochimica et Cosmochimica Acta* 12 (1957): 1–28.
2. I.A. Kunasz, *Lithium Raw Materials,* forthcoming.
3. Ibid.
4. D.E. White, "Thermal Waters of Volcanic Origin," *Geological Society of America Bulletin* 68 (1957): 1637–1658. E.J. Mayhew and E.B. Heylmun, "Complex Salt and Brines of the Paradox Basin," *Second Symposium on Salt,* ed. J.L. Rau, Cleveland: The Northern Ohio Geological Society, Inc. (1966): 221–235.
5. J.J. Norton, "Lithium, Cesium, and Rubidium—The Rare Alkali Metals," U.S. Geological Survey Professional Paper 820 (1973).
6. T.L. Kesler, "Exploration of the Kings Mountain Pegmatites," *Mining Engineering* 13 (1961): 1062–1068.
7. Kunasz, *Lithium Raw Materials.*

8. J.J. Norton and D.M. Schlegel, "Lithium Resources of North America," Bulletin 1027–G, U.S. Geological Survey (1955): 325–350. T.L. Kesler, "Lithium Raw Materials," in J.L. Gillson, ed., *Industrial Minerals and Rocks* (New York: AIME, 1960): 521–531.

9. F.B. Shay, "Low Cost Lithium from Brine at Silver Peak, Nevada," paper presented at AIME Annual Meeting, Los Angeles, California, February 21, 1967.

10. W.T. Barrett and B.J. O'Neill, Jr., "Recovery of Lithium from Saline Brines Using Solar Evaporation," in J.L. Rau and L.F. Dellwig, eds., *Third Salt Symposium* (Cleveland: The Northern Ohio Geological Society, Inc., 1970): 47–50.

11. Kunasz, *Lithium Raw Materials.*

12. Ibid.

13. Ibid.

14. Ibid.

15. "Klockner Buys SWA Lithium," *Metal Bulletin* 26 (April 5, 1968).

16. W.T. Gevers, private communications (1952–1953).

17. R. Mulligan, "The Geology of Canadian Lithium Pegmatites," Economic Report No. 21, Geological Survey of Canada (1965).

18. V. Angelleli and C.A. Rinaldi, "Resena Acerca de la Estructura, Mineralizacion, y Aprovechamento de Nuestras Pegmatita Portadoras de Minerales de Litio," *Acta Geologica Lilloana* 5 (1965): 1–18.

19. R.O. Roberts, "Amblygonite and Associated Minerals from the Mbale Mine, Uganda," *Imperial Institute Bulletin* 46 (1948): 342–347.

20. R.B. Ellestad and K.M. Leute, "Method of Extracting Lithium Values from Spodumene Ores," U.S. Patent 2,516,109 (1950).

21. M. Archambault, "Lithium Carbonate Production," U.S. Patent 3,112,171 (1963).

22. M. Archambault and C.A. Olivier, "Direct Production of Lithium Borate," U.S. Patent 3,112,168 (1963).

23. M. Archambault, C.A. Olivier, H.P. Lemay, and M. Savard, "Sodium-Ammonium Compounds Process for Extracting Lithium from Spodumene," U.S. Patent 3,112,170 (1963).

24. M. Archambault, C.A. Olivier, J.J. Panneton, and P. Fortier, "Production of Various Lithium Salts," U.S. Patent 3,112,172 (1963).

3
The Lithium Aluminosilicates

Apart from lithium carbonate, which is obtained from brines or by extraction from lithia minerals, the most important domestic source of lithia is spodumene. Spodumene is one of three naturally occurring lithium aluminosilicates, the other two being petalite and eucryptite. Figure 3.1 shows the location of these three natural lithium aluminosilicates on a ternary diagram. The location of metakaolin is also shown in this figure because mixtures of spodumene and kaolin and petalite and kaolin are widely used in refractory applications. Petalite is little used in the domestic ceramic industry at the present time because of the United Nations sanctions on Rhodesian raw materials. Eucryptite does not occur in sufficient quantities to be mined and sold as a concentrate. All domestic eucryptite is synthesized and sold in the beta form.

Hatch divided the $Li_2O.Al_2O_3$-SiO_2 system into five fields: the silica field, the field of beta spodumene–silica solid solution, the field of beta spodumene solid solution, the field of beta eucryptite solid solution, and the gamma alumina field. He reported that no compounds having the petalite (1:1:8) composition were stable above the solidus.[1]

The melting points of petalite, spodumene, and eucryptite are 1356°C, 1423°C, and 1397°C, respectively. However, it must be pointed out that commercial petalite and spodumene materials are concentrates, containing perhaps only 80–90% of the lithium mineral. The other minerals in the concentrates are typically feldspars, quartz, and micas. Since feldspars and micas form eutectics with spodumene and petalite, the melting points of the commercial concentrates are below the values indicated in the phase diagram. This is a point that should be recognized when using spodumene and petalite for both glass and refractory applications.

Although Hatch failed to synthesize the alpha forms of spodumene and petalite, he did recognize that high pressure was probably a necessary condition for the formation of alpha spodumene because of the large density difference between the alpha and beta forms. Roy et al. studied the phase equilibria at the

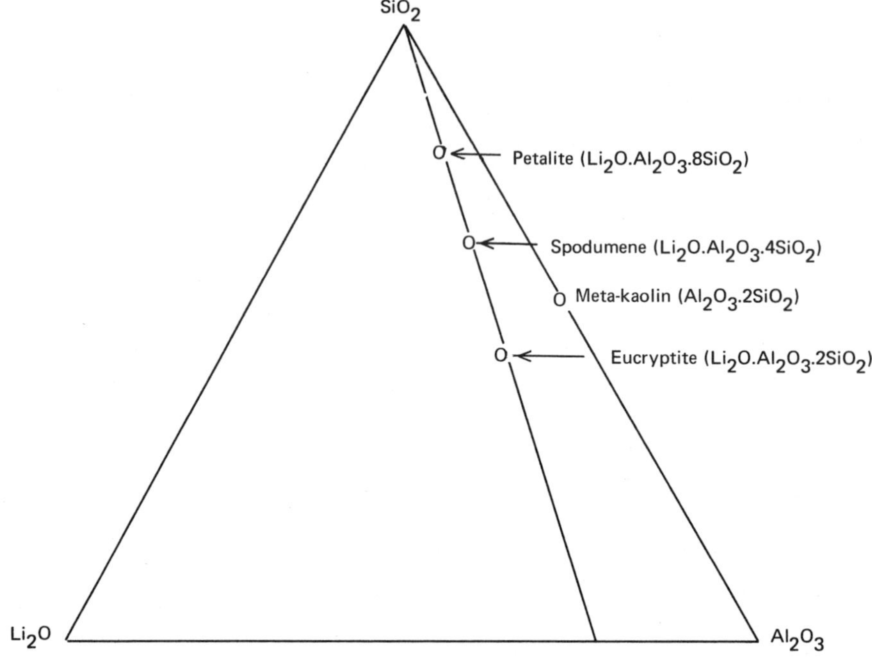

Figure 3.1. Location of Natural Lithium Minerals and Meta-kaolin

liquidus and at temperatures below the solidus along the join eucryptite-silica in the system Li_2O-Al_2O_3-SiO_2.[2] They succeeded in synthesizing the low or alpha forms of petalite, spodumene, and eucryptite by using a hydrothermal technique at high pressure. Petalite was reported to decompose congruently to beta spodumene solid solution at 680 ± 10°C, while spodumene inverted to beta spodumene at 500°C and a pressure of 10,000 psi. Eucryptite inverted at 972 ± 10°C to a high temperature form that melted incongruently. The alpha forms of these lithium aluminosilicates were also synthesized by Barrer and White.[3] They studied the formation of petalite, spodumene, and eucryptite in the temperature range 150–450°C by starting with a series of gels having the general formula $Li_2O.Al_2O_3nSiO_2mH_2O$, where n = 1–10. As a result of these studies, Barrer and White identified the species $Li_2O.Al_2O_3.2SiO_2.4H_2O$ and $Li_2O.Al_2O_3.8SiO_2.5H_2O$. These species showed the sorptive behavior of zeolites and underwent ion-exchange reactions involving Li^+, Ag^+, Na^+, K^+, Rb^+, Tl^+, NH_4^+, Ca^{++}, and Ba^{++}.

Skinner and Evans studied the crystal chemistry of beta spodumene solid solutions on the join $Li_2O.Al_2O_3$-SiO_2.[4] They showed that the stability field of beta spodumene solid solution is defined by the two-phase field beta spod-

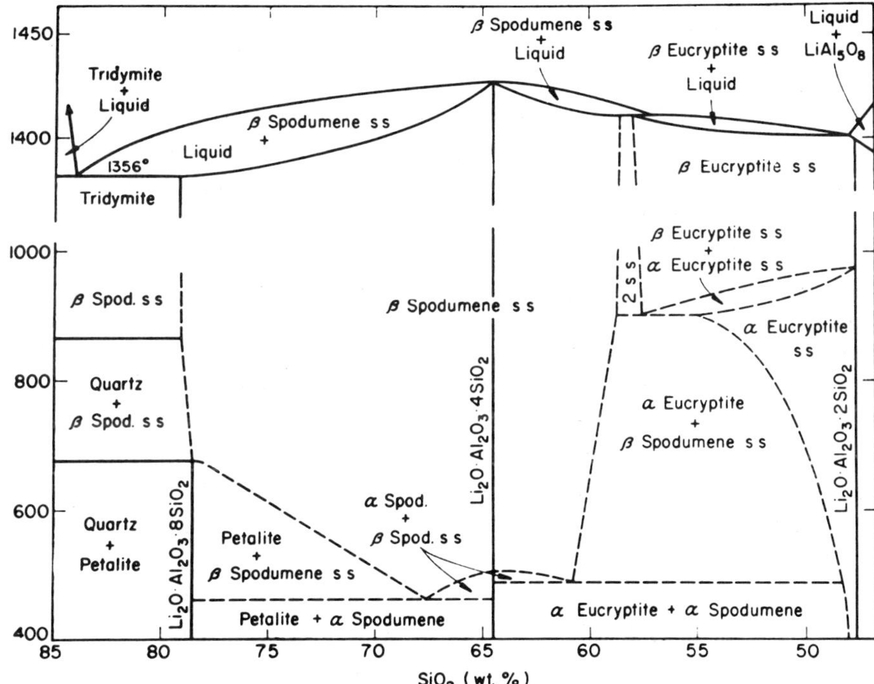

Figure 3.2. System $Li_2O.Al_2O_3.2SiO_2$ (eucryptite)-SiO_2. The high-temperature part of the diagram is after R.A. Hatch, *American Mineral* 28 (1943). The lower part represents conception at about 10,000 psi.

umene solid solution plus tridymite on the high silica side and by the two-phase field beta spodumene solid solution plus beta eucryptite solid solution on the low silica end. These findings are largely in agreement with those of Hatch and Roy et al.[5] Skinner and Evans, however, proved the existence of a narrow two-phase field of beta spodumene solid solution plus beta eucryptite solid solution. They confirmed that beta spodumene has a structure similar to that of keatite, which classes it with the so-called stuffed silica structures of Buerger.[6] In this concept, beta spodumene can be considered as being derived from the silica polymorph keatite by replacing some of the silicon atoms by aluminum atoms and stuffing an equal number of lithium ions into the interstices of the framework to maintain charge balance. Channels extend throughout the structure, providing paths along which lithium atoms can move. Skinner and Evans postulated that the existence of these channels endowed beta spodumene with cation exchange properties similar to those of the zeolites, although these may be restricted to Li^+ and H^+ ions.

Murthy and Hummel studied the phase equilibria in the system lithium metasilicate–beta eucryptite and found a eutectic at 1070°C and 57% eucryptite.[7]

The infrared spectra of compounds and solid solutions in the lithia-alumina-silica system were studied by Murthy and Kirby.[8] They demonstrated that Si^{4+} is replaced by $(Li^{1+}Al^{3+})$ on going from keatite to beta eucryptite. This substitution is evidently prevented initially because of the lack of success in making beta spodumene–keatite solid solutions containing more than 84.3% silica. The infrared spectra (after Murthy and Kirby) of beta spodumene–silica solid solutions and beta spodumene–beta eucryptite solid solutions are shown in Figures 3.3 and 3.4 The relative intensity of the 10 micron band increases as the silica content of the beta spodumene solid solution decreases. This band is the strongest for beta spodumene containing 64.6% silica. It is also strong in beta eucryptite (47.7% SiO_2). As Murthy and Kirby point out, the 10 micron band appears as soon as Al^{3+} enters the structure, and it intensifies as the Al^{3+} content increases. The conclusion is that this band is related to tetrahedrally coordinated Al^{3+} in the structures studied.

The structure of beta spodumene can be visualized best, perhaps, by examination of the stereographic projection along the *a* axis as shown by Ostertag et al.[9] This projection was plotted by a computer program using crystallographic data taken from Li and Peacor.[10]

In many applications, lithium behaves more like an alkaline earth element than an alkali. For example, the beta spodumene solid solution series has no counterpart in the other alkali systems, but it is similar to the solid solution series in the $MgO\text{-}Al_2O_3\text{-}SiO_2$ system. The phase equilibria in the $Li_2O\text{-}MgO\text{-}Al_2O_3\text{-}SiO_2$ system are, therefore, of some interest to ceramists. Prokopowicz and Hummel studied this system and indicate that both cordierite ($2MgO.2Al_2O_3.\text{-}5SiO_2$) and beta spodumene ($Li_2O.Al_2O_3.4SiO_2$) are the bases for low-expansion, thermal-shock-resistant bodies.[11] A quaternary glass containing 6% Li_2O, 4.5% MgO, 4.5% Al_2O_3, and 85% SiO_2 with an expansion of 51×10^{-7} (25–400°C) and a softening point of 550°C was reported. Because the solid solubility of MgO in beta spodumene and beta eucryptite is relatively low to almost negligible, the addition of magnesia to such bodies should result in only minor variations in the thermal expansion. Also, the effect of lithia additions to cordierite bodies can be explained by a mixture of two phases.

Karkhanavala and Hummel studied phase equilibria along the nonbinary join between cordierite and beta spodumene from 800°C to 1550°C using the quench technique.[12] They found the high-temperature phase relations between spodumene and cordierite to be very complex and to involve spinel, mullite, and corundum besides the two end members. The low-temperature

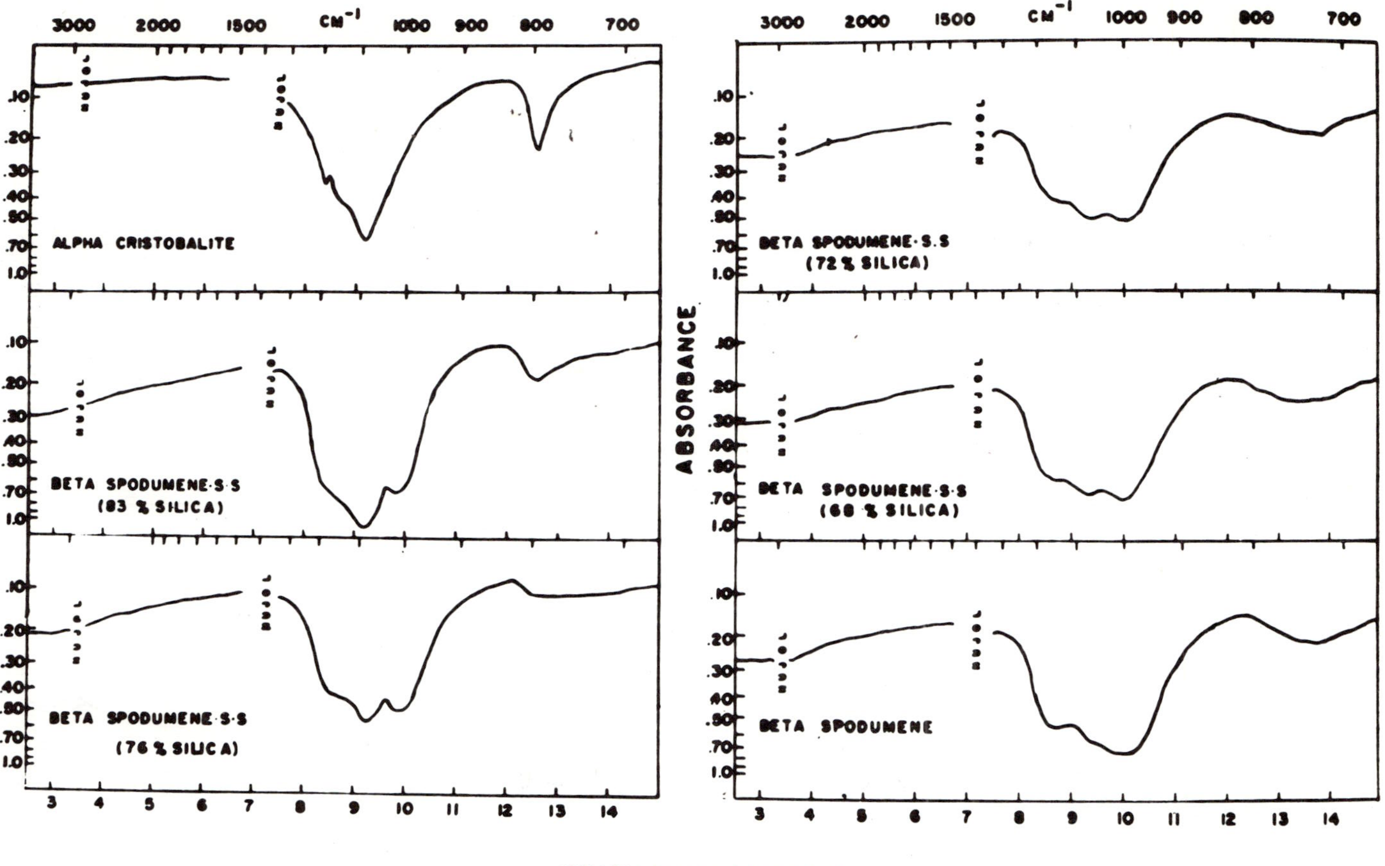

Figure 3.3. Infrared Spectra of Beta Spodumene–Silica Solid Solutions

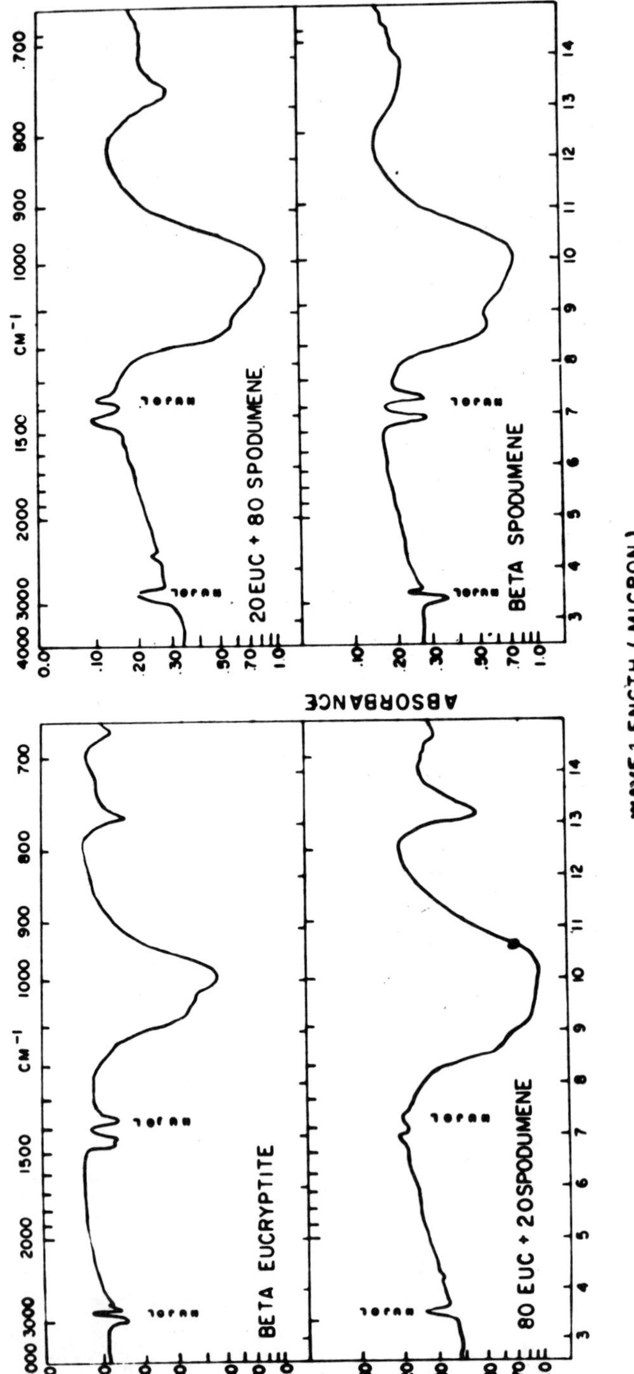

Figure 3.4. Infrared Spectra of Beta Spodumene–Eucryptite Solid Solutions

phase relations indicated a complete series of metastable solid solutions between μ-cordierite and beta spodumene. According to the authors, no improvement in thermal shock properties of the end member crystals can be expected by the reaction of alpha cordierite and beta spodumene at high temperatures. This is because of the development of a liquid at temperatures lower than the congruent melting point of spodumene or the incongruent melting point of cordierite and because of the development of high-expansion phases such as mullite and spinel.

Several investigators have studied ionic substitutions in the Li_2O-Al_2O_3-SiO_2 systems. For example, Ray studied the replacement of Li^+ by Mg^{2+}, and Al^{3+} by B^{3+} or Ga^{3+}, in the keatite lattice.[13] The crystallization and stability of keatite solid solutions of $Li_2O.Al_2O_3.nSiO_2$ were studied. Keatite solid solutions with greater than 40 mole percent Li_2O replaced by MgO (n = 3.7-5.7) were crystallized. The stability of the phase decreased with increasing MgO and/or SiO_2, and the metastable solid solution of this composition exsolved cordierite at higher temperatures. Stable keatite solid solutions with 25 mole percent of Al_2O_3 replaced by B_2O_3 (n = 3-5) were also crystallized. When n = 3, the lattice accommodated more Li^+ ions than it could without the replacement. Gallium sesquioxide can substitute for Al_2O_3 in a stable keatite lattice only for $n > 4$. When $n = 4$, $\geqslant$ 50 mole percent Al_2O_3 can be replaced by Ga_2O_3.

Tien and Hummel, in one of their reports on lithium oxide systems, reported that although the substitution of Ge^{4+} for Si^{4+} in the alpha (phenacite structure) and beta (beta quartz structure) forms of eucryptite increased the a and c lattice parameters at room temperature, the axial thermal expansion coefficients for the solid solution phases are practically the same as for the unsubstituted phases.[14]

Hummel et al. synthesized the chromium analogue of alpha spodumene and studied its solubility relationships with ureyite ($NaCrSi_2O_6$) and beta spodumene.[15] The chromium analogue, $LiCrSi_2O_6$, was synthesized from chemically pure materials at 1000-1200°C. The inclusion of Cr in the beta spodumene structure had little effect on its low linear thermal expansion coefficient. The linear expansion coefficient of the alpha-chrome-spodumene was 62×10^{-7} in the range 85°C to 1000°C.

Thermal Expansion

Perhaps the most unusual property of the lithium aluminosilicates is the remarkably low expansion coefficients exhibited by the beta forms of the minerals. As shown in Figure 3.5, these vary from slightly positive, for beta spodumene, to highly negative, for beta eucryptite. These low-expansion

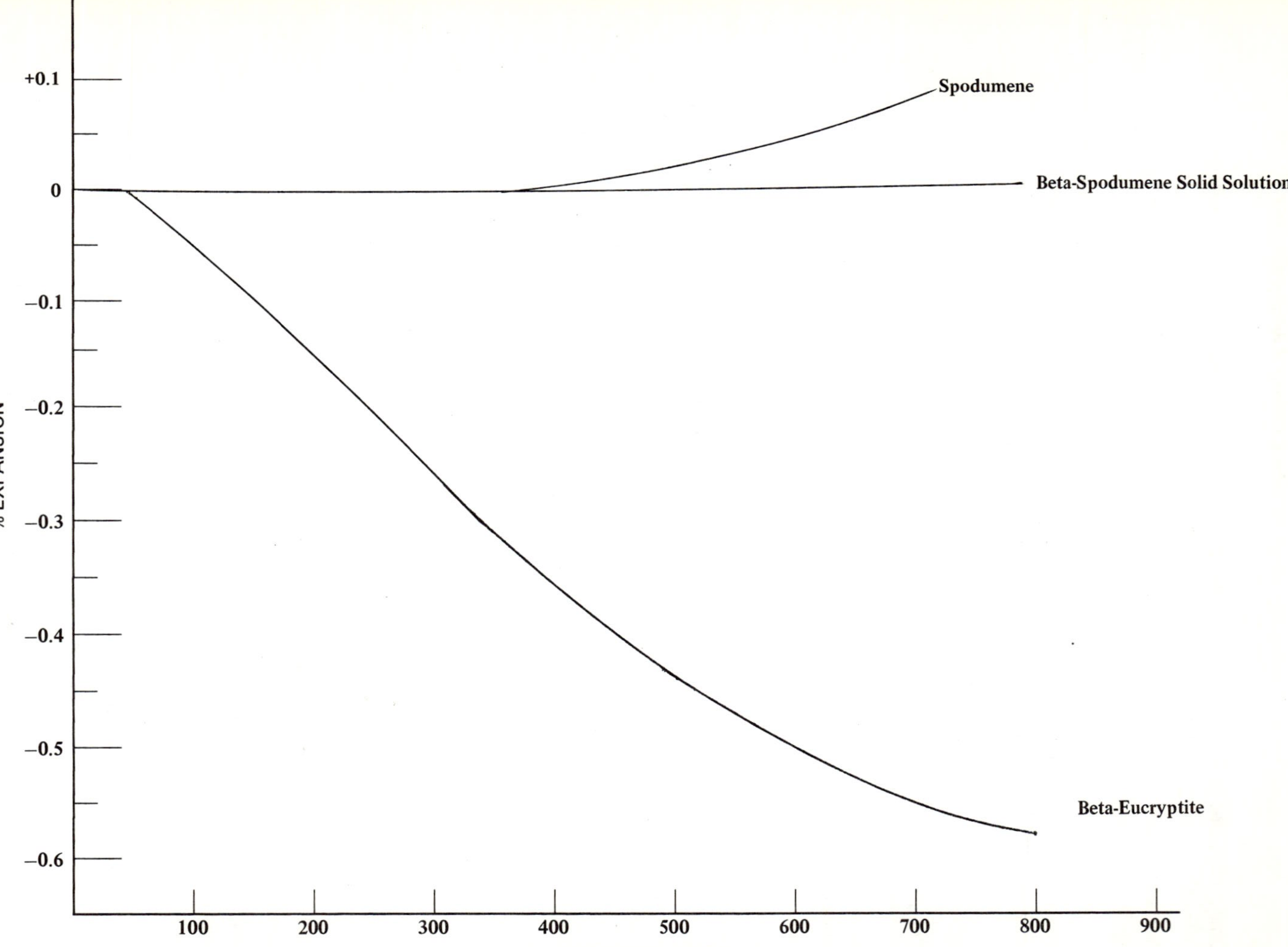

Figure 3.5. The Thermal Expansion Curves for Beta Spodumene, Beta Spodumene Solid Solution (Calcined Petalite), and

coefficients have been used to advantage in the formulation of thermal-shock-resistant refractories, which are discussed in a later chapter. The thermal contraction of beta eucryptite is of value for counteracting the high expansion of epoxy resins in such applications as the encapsulation of electrical motors.

Hummel was the first to report the unusually low thermal expansion coefficient of beta spodumene, which is made by calcining natural or alpha spodumene above its transformation point (see p. 40). [16] He reported the thermal expansion of alpha spodumene to be $4 \times 10^{-6}/°C$ up to about 900°C, when it transformed to beta spodumene with a linear expansion of $1.9 \times 10^{-6}/°C$. After firing beta spodumene bodies to 1100°C and 1200°C, the thermal expansion increased slightly to $2.4 \times 10^{-6}/°C$ due to the formation of a glassy phase. More recent thermal expansion data on beta spodumene obtained by the author indicate that the linear thermal expansion of beta spodumene is about $1 \times 10^{-6}/°C$ over the range room temperature to 900°C, or roughly half of the value reported by Hummel. The expansion of calcined petalite, or beta spodumene solid solution, was reported by Hummel to be about the same as or slightly lower than that of fused silica ($0.5 \times 10^{-6}/°C$).

The thermal expansion characteristics of synthetic lithium aluminosilicate bodies are usually somewhat different from those of the natural minerals. This is undoubtedly because, as previously mentioned, the natural materials are concentrates that contain other mineral species, while synthetic minerals are normally made to stoichiometric proportions. Hummel also studied the thermal expansion properties of synthetic lithia aluminosilicates with ratios (Li_2O:Al_2O_3:SiO_2) from 1:1:2 to 1:1:15.[17] He noted that there is a progressive shift in the d spacing of the most intense x-ray line (d = 3.47 to 3.44) as more and more silica is incorporated into the solid solution. The following thermal expansion coefficients were reported for the range 0 to 1000°C: 1:1:4 (spodumene) = $0.9 \times 10^{-6}/°C$; 1:1:6 = $0.5 \times 10^{-6}/°C$; 1:1:8 = $0.3 \times 10^{-6}/°C$; and 1:1:10 = $0.5 \times 10^{-6}/°C$. In the last sample (1:1:10), the presence of quartz and cristobalite indicates the amount of silica that the beta spodumene crystal lattice can accommodate. An interesting phenomenon was noted in the behavior of the quartz inversion in the thermal analysis of samples 1:1:10 and 1:1:12. Evidently, the presence of lithia caused the transformation to take place at 400°C and 500°C, instead of the usual 573°C. The large thermal contraction of beta eucryptite was noted, and the structure was compared with beryl and high quartz as being examples of hexagonal spiral or ring structures which exhibit either low expansions or contractions.

In another paper, Hummel discussed the practical aspects of the thermal contraction of beta eucryptite.[18] As he points out, ice I and bismuth have long been known to have a lower density than liquid phases of the same composition; beta eucryptite also floats on its liquid and shows a continuous thermal contraction from room temperature to 1000°C. The following practical applications for the thermal contraction were listed as:

1. Use of beta eucryptite as a new thermal expansion standard providing expansion coefficients from negative to positive.
2. The formulation of mechanical mixtures of crystals to provide a material with a zero coefficient of expansion. For example, Figure 3.6 shows the expansion characteristics of a mechanical mixture of corundum and eucryptite. Although such a body would exhibit a zero coefficient in a dilatometer, it would be basically unsound for thermal shock resistance because it provides the maximum possible movement of one crystal phase relative to the other.
3. Use of beta eucryptite in glasses to alter the expansion coefficient. In this case, the eucryptite can be mechanically mixed with the glass or devitrified

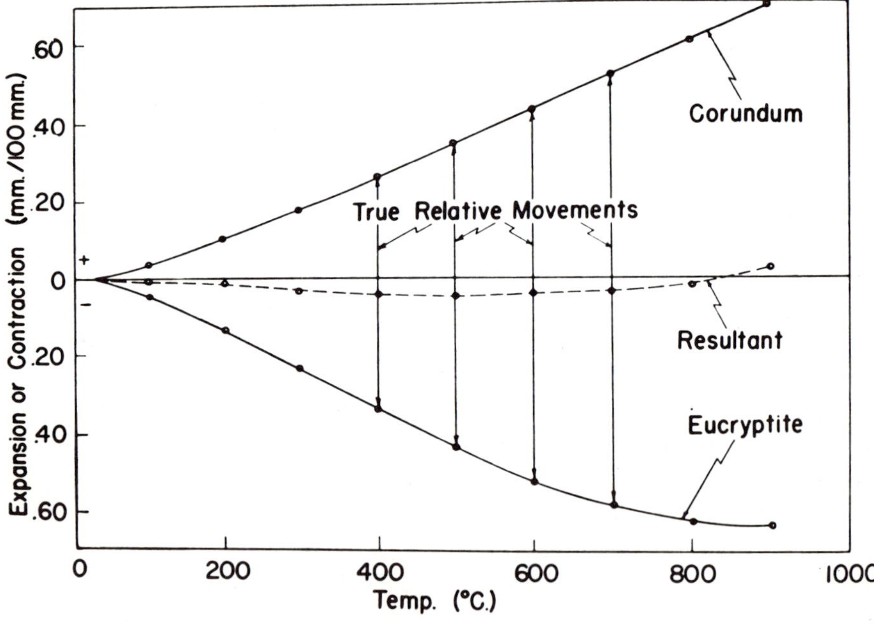

Figure 3.6. Expansion Characteristics of a Mechanical Mixture of Corundum and Eucryptite. *Source:* F. A. Hummel, "Significant Aspects of Certain Ternary Compounds and Solid Solutions," *Journal of the American Ceramic Society* 35 (1952): 64–66.

in situ. Here, Hummel correctly predicted the recent developments in glass-ceramics, such as the use of large-dimension telescope mirrors and low-expansion cookware.

Smoke synthesized various lithium aluminosilicate compositions mainly with a view to making bodies that would withstand severe thermal shock. He identified two areas in which the linear thermal expansion is negative.[19] These are shown in Figure 3.7. The linear thermal expansions range from 0 to -0.38% for the area in which beta eucryptite solid solution is the principal crystalline phase, and from 0 to -0.04% for the area in which beta spodumene solid solution is the main phase. Some typical thermal expansion curves for lithium aluminosilicate bodies are compared with fused quartz in Figure 3.8. In this figure, QAl, QA5, Q, and QA4 refer to the following respective compositions: 100% spodumene; 6% Li_2O, 22.9% Al_2O_3, 71.1% SiO_2; petalite; and 4.6% Li_2O, 17.6% Al_2O_3, 77.8% SiO_2. It is interesting to compare the thermal expansion curve for synthetic spodumene (shown in Figure 3.8) with that for beta spodumene made from the mineral concentrate (shown in Figure 3.5). While the natural spodumene gives an expansion curve that remains positive at all temperatures, the synthetic product gives a curve that shows some thermal contraction up to 400°C. As previously mentioned, this is undoubtedly due to the fact that the synthetic product is made to stoichiometric proportions, while the natural product is a relatively impure concentrate containing other mineral species.

White and Rigby synthesized two series of mixes with compositions varying from 1:1:2 ($Li_2O:Al_2O_3:SiO_2$) to 1:1:8.[20] In the first series, pure silica and hydrated alumina were mixed with lithium carbonate in the correct molecular proportions. In the second series, the compounds were synthesized from ball clay, silica, and lithium carbonate. In all cases, the bodies were fired at 1200°C in a gas-fired furnace. The reversible thermal expansion coefficients for the 1:1:2, 1:1:4, 1:1:6, and 1:1:8 compounds in the range 20°C to 1000°C were reported as follows (series I and series II respectively): 0.41 and 1.43; 2.04 and 1.63; 0.41 and 0.92; and 1.63 and $0.92 \times 10^{-6}/°C$. In series I, negative expansions were exhibited by the 1:1:2 and 1:1:6 compositions. The former contracted up to a temperature of 835°C, when it regained its original length; the maximum negative reading recorded was $-7.8 \times 10^{-6}/°C$ between 600°C and 650°C. For the 1:1:6 composition, the maximum negative reading was $-2 \times 10^{-6}/°C$ between 350°C and 375°C. The 1:1:4 and 1:1:8 compositions exhibited positive expansions at all temperatures. In series II, none of the compositions showed negative expansions.

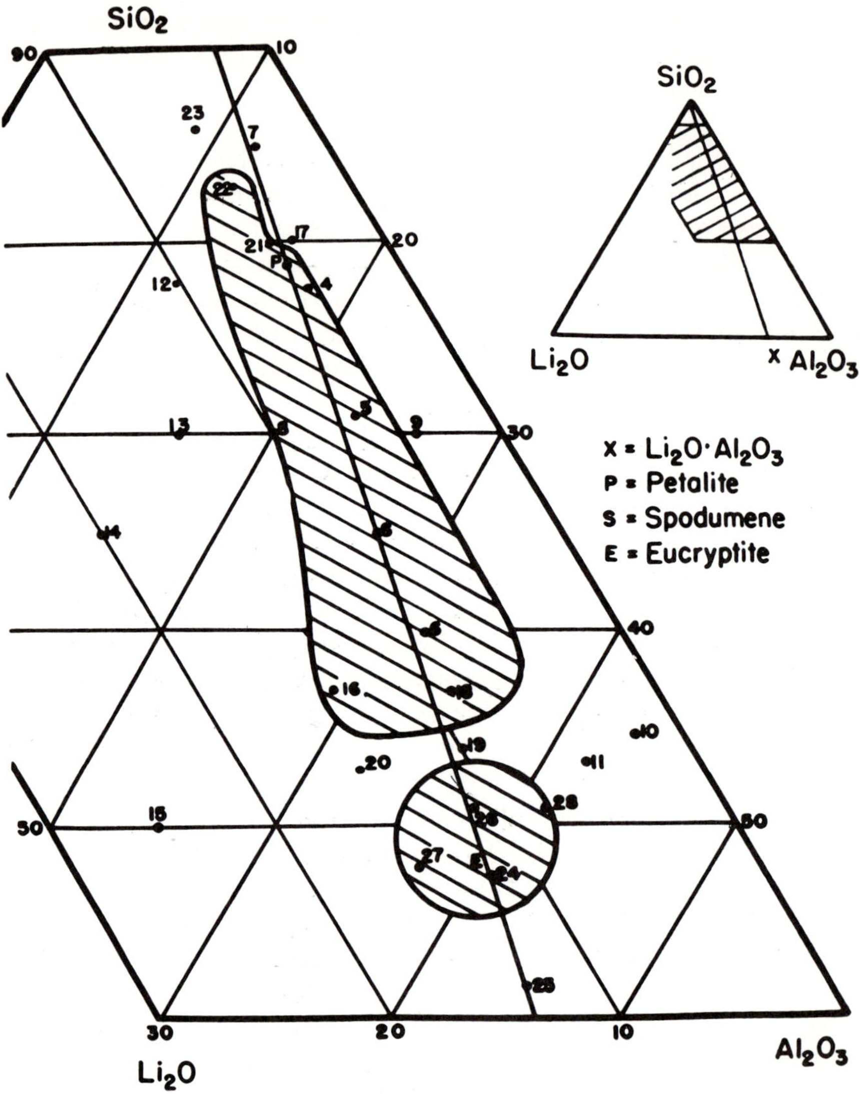

Figure 3.7. Areas of Negative Linear Thermal Expansion in the System Li_2O-Al_2O_3-SiO_2, Q Series. *Source:* E.J. Smoke, "Ceramic Compositions Having Negative Linear Thermal Expansion," *Journal of the American Ceramic Society* 34 (1951): 87–90.

Another batch of materials representing both series I and II was prepared, but this time they were fired in an electric kiln at 1200°C. The thermal expansion coefficients of the 1:1:2, 1:1:4, 1:1:6, and 1:1:8 compositions in the

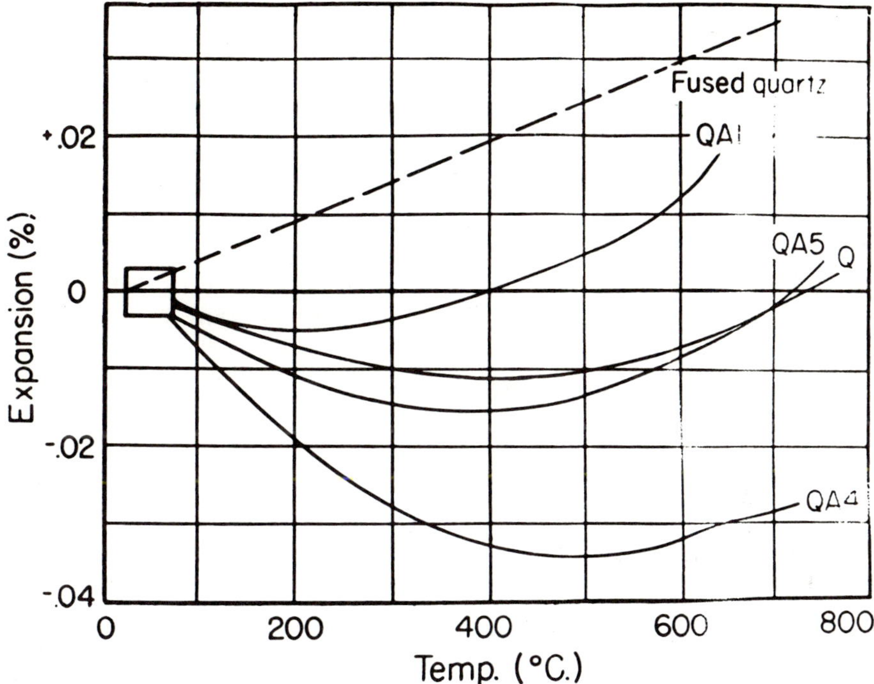

Figure 3.8. Linear Thermal Expansion of Mineral and Synthetic Compositions. *Source:* Ibid.

range 20°C to 1000°C were given as: 1.22 and 1.73; 2.45 and 1.43; 1.84 and 1.12; and 2.96 and 1.02 × 10^{-6}/°C. Whereas negative expansions had been obtained with certain mixes in the gas-fired furnace, all electrically fired compositions exhibited positive expansions.

The variation in thermal expansion coefficients exhibited by different mixtures in the Li_2O-Al_2O_3-SiO_2 system is presumably largely due to the differing degree of equilibrium attained by each mixture. Thus, the presence of small amounts of beta eucryptite, on the one hand, or cristobalite, on the other, will have a significant effect on the overall thermal expansion.

Bulavin and Medvedovskaya investigated the physicochemical processes that occur during the synthesis of various lithium aluminosilicates from lithium carbonate, kaolin, and quartz sand.[21] They showed that beta spodumene is formed at 1150–1250°C as a result the following reaction between eucryptite and quartz:

$$Li_2O.Al_2O_3.2SiO_2 + 2SiO_2 \longrightarrow Li_2O.Al_2O_3.4SiO_2.$$

At 1350°C, beta eucryptite transforms to beta spodumene and lithium aluminate according to the following reaction:

$$(2Li_2O.Al_2O_3.2SiO_2) \longrightarrow Li_2O.Al_2O_3.4SiO_2 + Li_2O.Al_2O_3$$

All of the thermal expansion coefficients reported above by various investigators are average results obtained from polycrystalline materials. Gillery and Bush studied the mechanism that causes the thermal contraction of beta eucryptite and concluded that the *a* axis of the unit cell expands during heating and the *c* axis contracts.[22] This is shown graphically in Figure 3.9, which shows the axial and aggregate expansions of beta eucryptite. The measured coefficients were -17.6×10^{-6} parallel to the *c* axis and $+8.21 \times 10^{-6}$ perpendicular to the *c* axis.

The structure of beta eucryptite consists of Si-O tetrahedra spirals, which can be compared with the springs of a spring mattress. The spirals are fixed in such a manner that, if the material is heated, the spiral is put under a torsional stress. Because of the extreme thermal-expansion anisotropy of beta eucryptite,

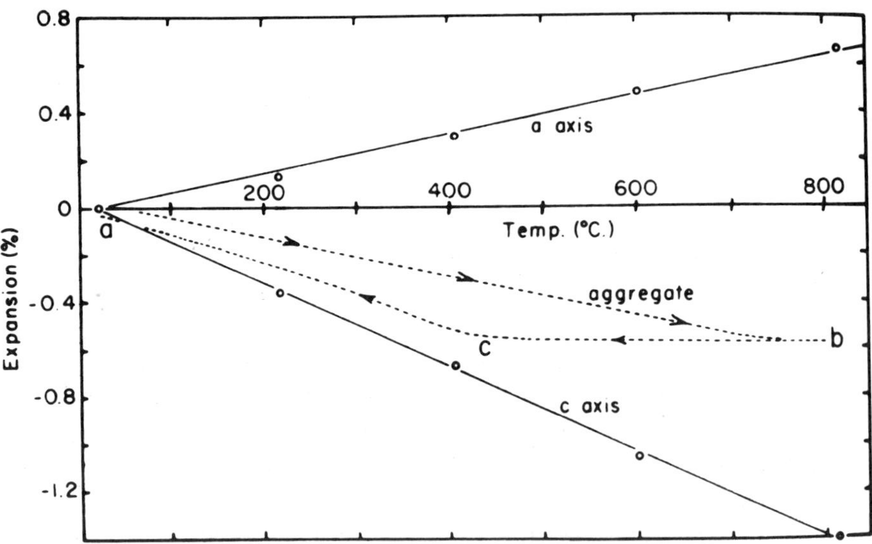

Figure 3.9. Axial and Aggregate Expansions of Beta Eucryptite. *Source:* F. H. Gillery and E. A. Bush, "Thermal Contraction of Beta Eucryptite ($Li_2O.Al_2O_3.2SiO_2$) by X-Ray and Dilatometer Methods," *Journal of the American Ceramic Society* 42 (1959).

bodies made from it will exhibit high internal stresses. Bush and Hummel have shown that the resulting fractures heal during reheating.[23]

A similar thermal-expansion anisotropy is exhibited by beta spodumene solid solutions. Ostertag et al. demonstrated that the anisotropy increases slightly and the volume expansion decreases as the compositions become richer in silica.[24] The thermal expansions of beta spodumene solid solutions ($Li_2O.Al_2O_3nSiO_2$, for n = 4–9) are given in Table 3.1 and the linear thermal expansion along the *a* and *c* axes are shown in Figure 3.10. The lattice parameters *a* and *c* of the beta spodumene solid solutions decrease continuously throughout the solubility range as shown by the data in Table 3.2.

During the research investigation of low iron spodumene, in which lattice iron was removed by high-temperature chlorination, the author found that lithia could also be removed from the lattice under certain conditions.[25] It then became of interest to determine how much lithia could be removed from the lattice without altering the beta spodumene structure and the concomitant low thermal expansion. A graph showing the linear coefficient of thermal expansion of beta spodumene as a function of the lithia percentage is shown in Figure 3.11. High-temperature chlorination removed lithia down to about 2.3% Li_2O without affecting either the beta spodumene structure or the low thermal expansion coefficient. At 1.67% Li_2O and below, only mullite and cristobalite were found in the samples. The refractoriness (PCE values) decreased slightly from cone 15 to cone 14 at 2 o'clock as the percent lithia was reduced from 6.09 to 3.79 and then remained constant down to 2.06% Li_2O. As the lithia was further reduced to 0.99% Li_2O, the PCE value exceeded cone 19.

Table 3.1
Thermal Expansion of Beta Spodumenes along *a* and *c* Axes of $Li_2O.Al_2O_3.nSiO_2$

	Average Coefficient of Thermal Expansion × 10^{-8}					
Temperature (°C)	*n = 4*	*n = 5*	*n = 6*	*n = 7*	*n = 8*	*n = 9*
	Along *a* Axis					
25–300	–208	–253	–283	–299	–315	–334
25–600	–206	–250	–282	–298	–313	–332
25–900	–204	–245	–275	–289	–304	–320
15–1200	–202	–241	–268	–281	–297	–311
	Along *c* Axis					
25–1100	+645	+643	+630	+626	+621	+565

Source: W. Ostertag et al., "Stereographic Projection of Crystal Structures: Example, Beta Spodumene," *Journal of the American Ceramic Society* 52 (1969).

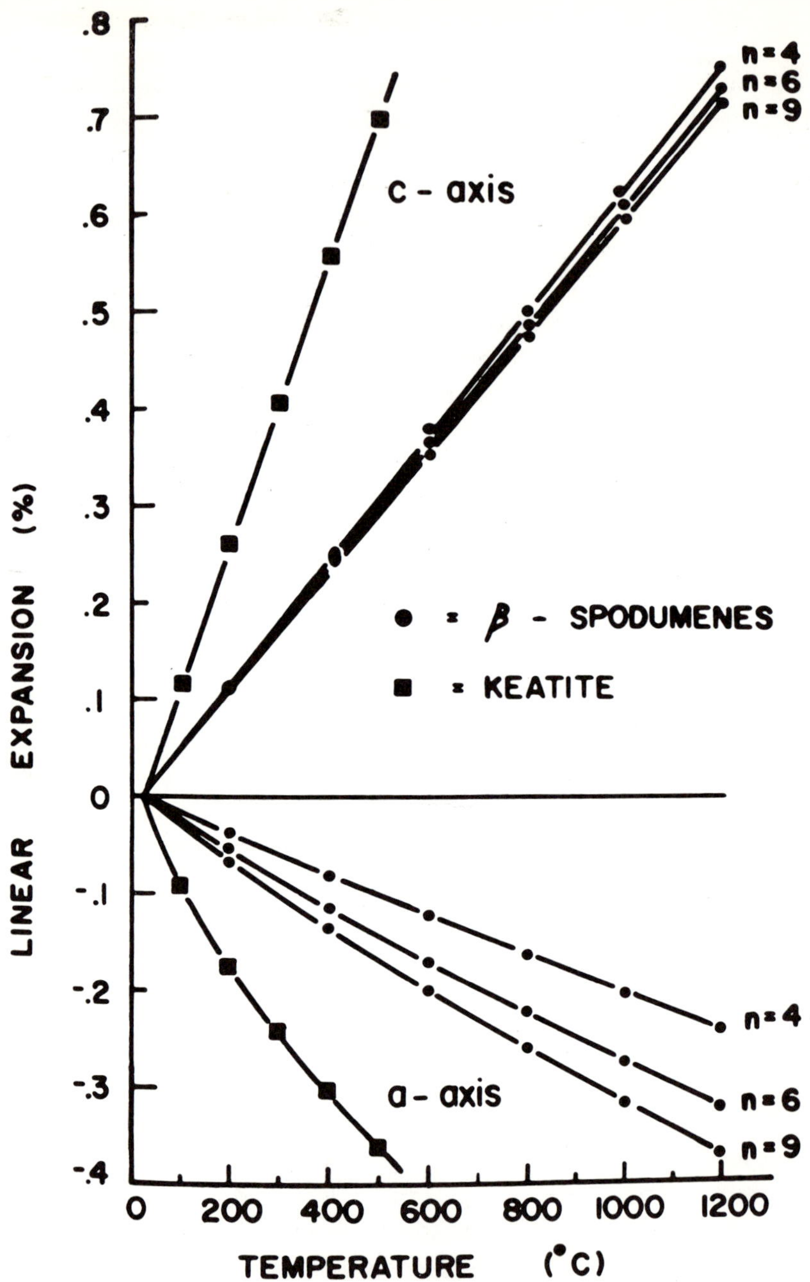

Figure 3.10. Linear Thermal Expansion in Direction of *a* and *c* Axes in $Li_2O.Al_2O_3.nSiO_2$. *Source:* W. Ostertag et al., "Thermal Expansion of Synthetic Beta Spodumene and Beta Spodumene-Silica Solid Solutions," *Journal of the American Ceramic Society* 51 (1968): 651-654.

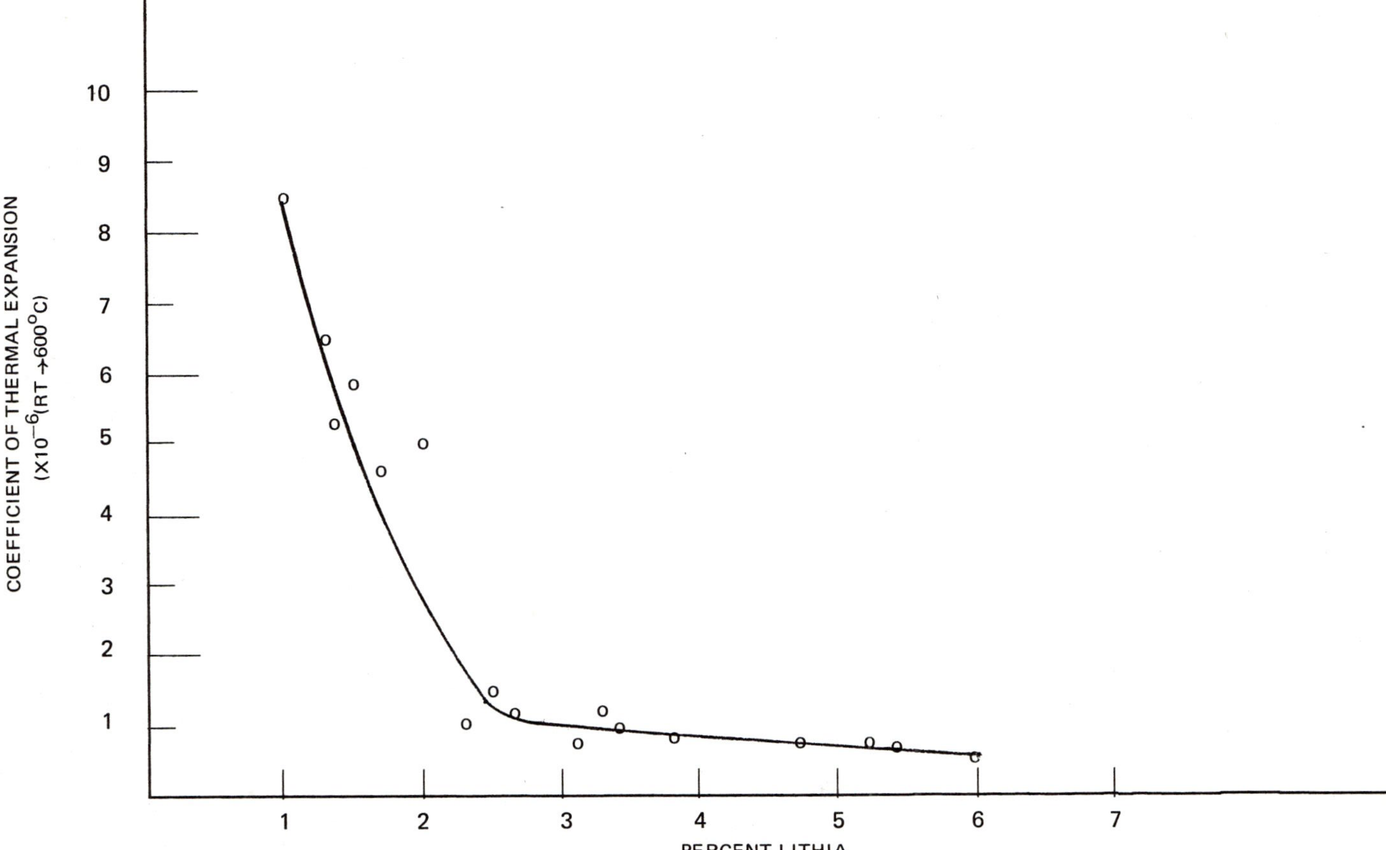

Figure 3.11. Coefficient of Thermal Expansion as a Function of Percent Lithia

Table 3.2
Lattice Parameters of Beta Spodumenes

Composition		*Lattice Parameters*		
Wt. % SiO_2	*Molar Ratio* $Li_2O.Al_2O_3.nSiO_2$	a_0 *(A)* ±*0.0008*	c_0 *(A)* ±*0.0008*	*Volume of Unit Cell* *(A^3)*
64.58	1:1:4	7.5378	9.1556	520.21
67.25	1:1:4.5	7.5316	9.1340	518.13
69.52	1:1:5	7.5263	9.1167	516.41
71.48	1:1:5.5	7.5209	9.1008	514.78
73.22	1:1:6	7.5158	9.0857	513.23
74.76	1:1:6.5	7.5104	9.0712	511.67
76.13	1:1:7	7.5055	9.0591	510.32
77.37	1:1:7.5	7.5003	9.0482	509.00
78.48	1:1:8	7.4962	9.0381	507.88
79.48	1:1:8.5	7.4925	9.0280	506.81
80.40	1:1:9	7.4892	9.0183	505.82

Source: Ostertag et al., "Stereographic Projection."

The Spodumene Transformation

All three naturally occurring lithium aluminosilicates undergo irreversible phase transformations. Since eucryptite is only made and used in the beta form and since the volume change associated with the petalite transformation is negligible, the spodumene phase change remains the most important of the three. Here, alpha spodumene is converted from a monoclinic pyroxene mineral to a less dense tetragonal structure. The specific gravity changes from about 3.2 to about 2.4. A suitable method for determining the extent of the transformation is to use a disappearing x-ray line technique. Here, the relative heights of the lines at $d = 4.21$ and $d = 2.93$ are measured and compared to standard samples.

Between 1901 and 1903, Doelter determined the melting point of spodumene as 1080–1090°.[26] He presumably confused the phase transformation with melting. Later information on the spodumene transformation was provided by Endell and Riecke [27] and Ballo and Dittler.[28] Endell and Riecke found a specific gravity change from 3.147 (at 20°) to 2.367 (at 1380°). A six-hour heating at 980° was sufficient to cause the transformation. The average refractive index changed from 1.66 to 1.519. The region of specific gravity change was thus defined at 920–980°C but identified as a region of melting. Ballo and Dittler correctly identified the specific gravity alteration as a phase change. They demonstrated that there is a 90% conversion to the beta form after heating at 1050° for 1.75 hours and after heating at 1200° for 0.5 hours.

Spodumene has been reported to invert to the high temperature polymorph about 720°C,[29] but Roy, Roy, and Osborn synthesized beta spodumene at 500°C and 10,000 psi H_2O pressure.[30]

Urazov et al. studied the rate of the phase transformation of three samples of spodumene containing 5.12%, 4.70%, and 6.05% Li_2O.[31] They used heating rates of 10 to 22 degrees per minute and found transformation temperatures varying from 990°C to 1100°C. As the heating rate was increased, the temperature at which the transformation occurred was increased and the range was broadened. They also studied the effect of additions of SiO_2 and K_2SO_4 on the temperature of the phase change at a constant heating rate of 11 degrees per minute. These additives lowered the temperature of the transformation, silica being more effective than potassium sulfate.

The temperature at which the transformation takes place is also affected by the presence of lattice impurities in the spodumene mineral.[32] Endell and Riecke also noted that very pure samples of spodumene (7.62% Li_2O) gave lower temperatures for the monotropic transformation.[33]

This author has studied the kinetics and thermodynamics of the spodumene transformation using a sample of spodumene (7.5% Li_2O) separated by heavy liquid techniques (methylene iodide to remove heavy minerals and acetylene tetrabromide to remove feldspar, mica, and quartz). The rate of the transformation, shown as the percent of alpha spodumene versus time at three temperatures, is presented in Figure 3.12. The transformation followed first order kinetics with an activation energy of 156 *k* cal.mole^{-1} or 420 cal/g. (see Figure 3.13). The heat of transformation (ΔH_{trans}) was measured using a high temperature DTA technique in which the area under the endothermic peak (971°C or 1780°F) was compared to that of a standard material. When the dissociation of calcium carbonate was used as a standard heat reference, a value of 24 cal/g. was found for $\Delta H_{trans.}$; when the heat of fusion of barium chloride was taken as the reference, a value of 29.6 cal/g. was obtained. Barany and Adami studied the heats of formation from the oxides at 298.15°K for alpha spodumene and beta spodumene.[34] Values of -21.0 ± 0.7 and -14.3 ± 0.7 *k* cal/mole respectively were reported. This suggests a much smaller value for $\Delta H_{trans.}$ than that found by the present author.

There are several practical examples of the use of the volume change associated with the spodumene transformation. The fact that the change is accompanied by decrepitation of the spodumene into a fine powder has been used to advantage in beneficiating the mineral.[35] Smoke has used the volume

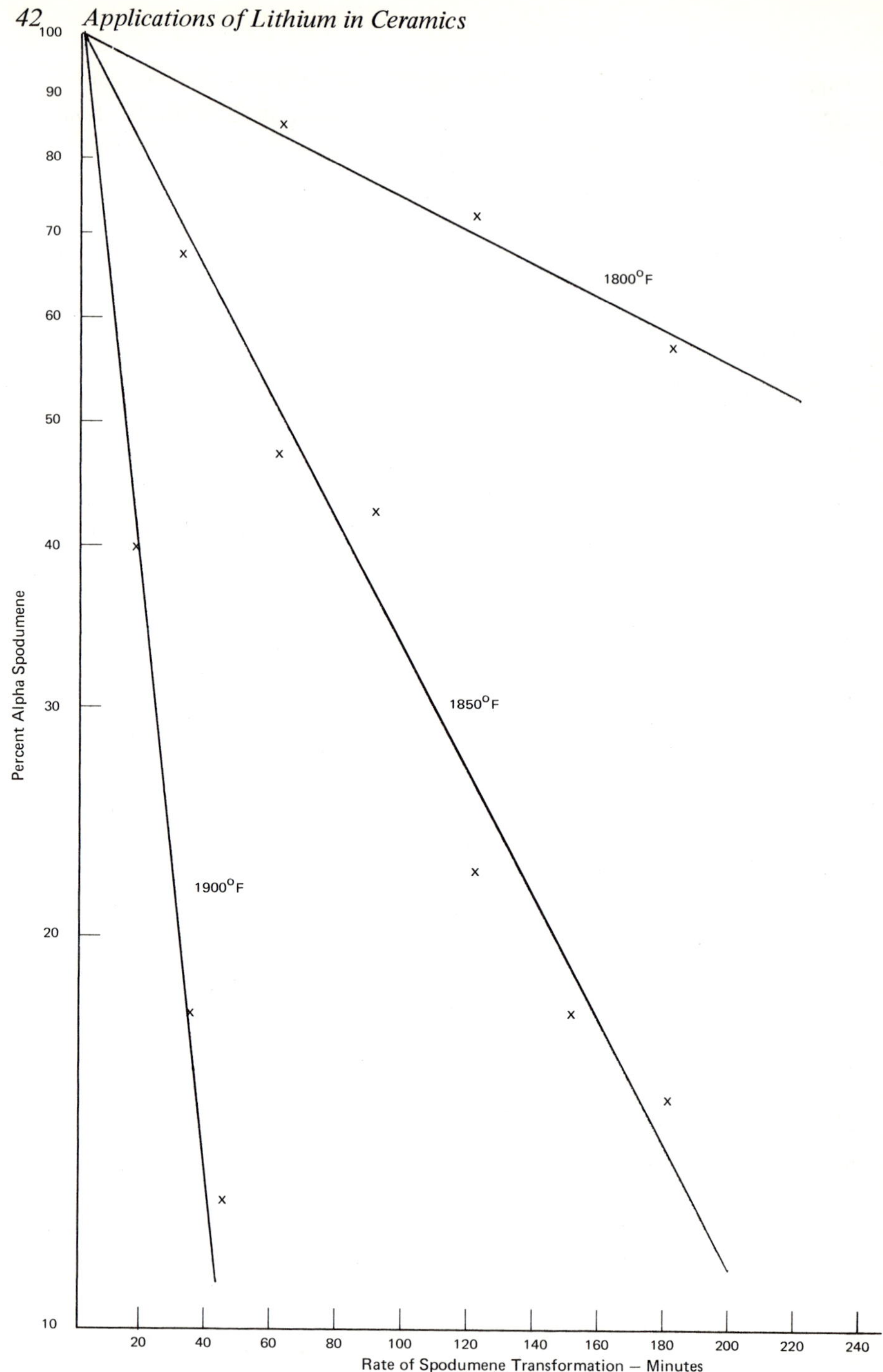

Figure 3.12. Rate of Spodumene Transformation as a Function of Temperature

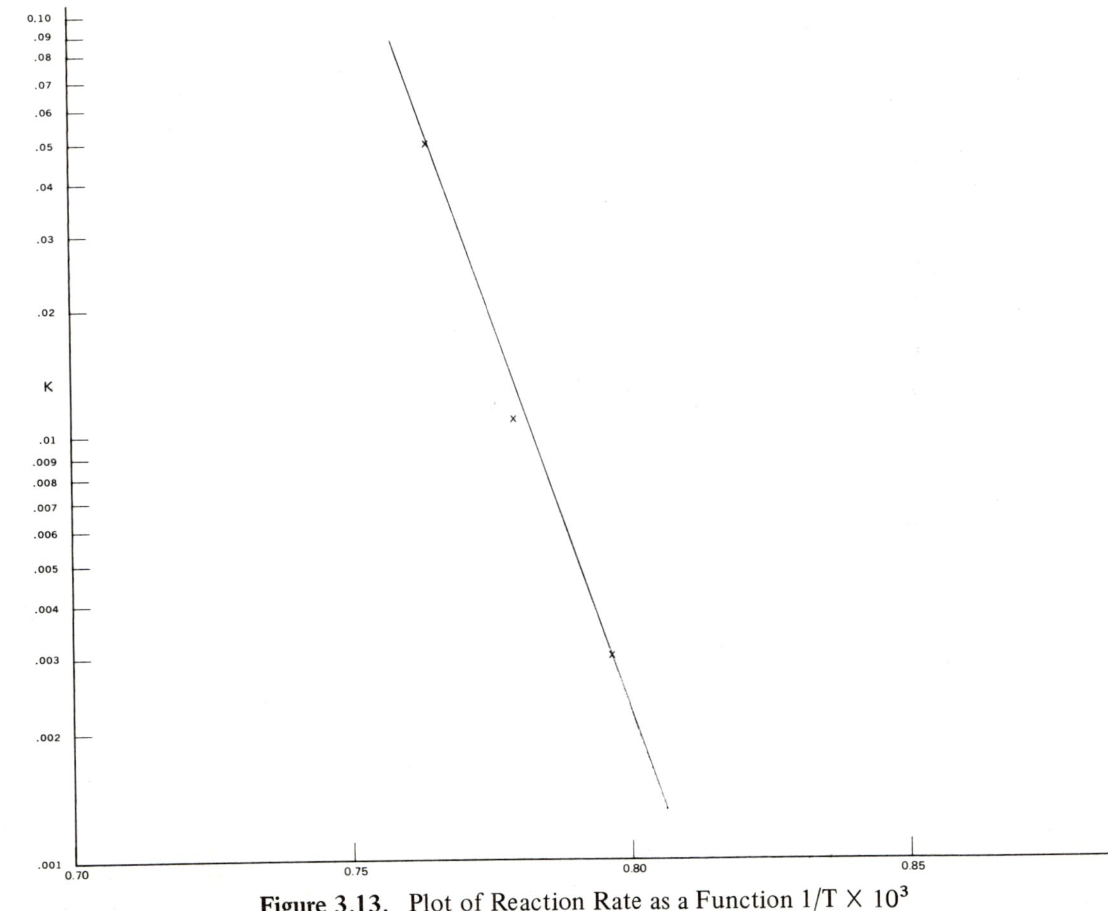

Figure 3.13. Plot of Reaction Rate as a Function $1/T \times 10^3$

increase to counteract the shrinkage of ceramic bodies.[36] He developed a body containing 60% spodumene and 40% lead bisilicate that exhibited a shrinkage of only 0.1% after firing at 2020°F. The shrinkage curve for this body is shown in Figure 3.14. Finally, he has used small amounts of alpha spodumene (5%–10%) to counteract the shrinkage of low-expansion refractories based on beta spodumene.

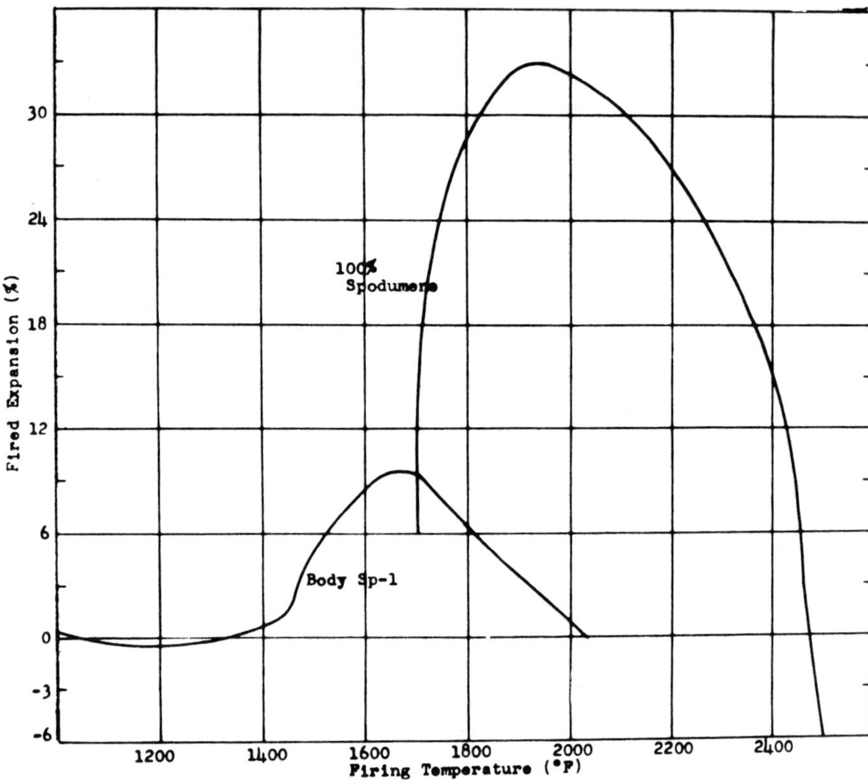

Figure 3.14. Firing Shrinkage versus Temperature of 100% Spodumene and Body Sp-l (60% Spodumene, 40% Lead Bisilicate. *Source:* E. J. Smoke, "Inorganic Dielectrics Research," *Rutgers University Engineering Research Bulletin,* no. 50 (1970): 82.

Summary

There are three naturally occurring lithium aluminosilicates: petalite, spodumene, and eucryptite. All undergo irreversible phase transformation to modifications that exhibit negative or small, but positive, thermal expansion coefficients. These negative or low expansion coefficients are the result of a structure that has been compared to a spring mattress, in which the spirals are placed under a torsional stress on heating. Of these transformations, the spodumene phase change is the most important because of the large volume change associated with it. Practical use has been made of this phase change for beneficiating the mineral and for counteracting shrinkage in ceramic bodies.

Notes

1. R.A. Hatch, "Phase Equilibrium in the System: $Li_2O.Al_2O_3$-SiO_2," *American Mineralogist* 28 (1943): 471–96.

2. R. Roy, D.M. Roy, and E.F. Osborn, "Compositional and Stability Relationships Among the Lithium Aluminosilicates: Eucryptite, Spodumene, and Petalite," *Journal of the American Ceramic Society* 33 (1950): 152–159.

3. R.M. Barrer and E.A.D. White, "The Hydrothermal Chemistry of Silicates. Part I, Synthetic Lithium Aluminosilicates," *Journal of the Chemical Society* (May 1951): 1267–1278.

4. B.J. Skinner and H.T. Evans, Jr., "Crystal Chemistry of Beta Spodumene Solid Solutions on the Join $Li_2O.Al_2O_3$-SiO_2," *American Journal of Science,* Bradley Volume 258-A (1960): 312–324.

5. Hatch, "Phase Equilibrium," Roy et al., "Compositional and Stability Relationships."

6. M.J. Buerger, "The Stuffed Derivatives of the Silica Structures," Amer. Min., 39 (1954): 600–614.

7. M.K. Murthy and F.A. Hummel, "Phase Equilibria in the System Lithium Metasilicate–Beta Eucryptite," *Journal of the American Ceramic Society* 37 (1954): 14–17.

8. M.K. Murthy and E.M. Kirby, "Infrared Study of Compounds and Solid Solutions in the System Lithia-Alumina-Silica," *Journal of the American Ceramic Society* 45 (1962): 324–329.

9. W. Ostertag, E.A. Hoar, and G.R. Fischer, "Stereographic Projection of Crystal Structures: Example, Beta Spodumene," *Journal of the American Ceramic Society* 52 (1969): 224.

10. C.T. Li and D.R. Peacor, "Crystal Structure of $LiAlSi_2O_6$–II (Beta Spodumene)," *Zeitschrift fur Kristallographie* 126 (1967): 46–65.

11. T.I. Prokopowicz and F.A. Hummel, "Reactions in the System Li_2O-MgO-Al_2O_3-SiO_2: II, Phase Equilibria in the High Silica Region," *Journal of the American Ceramic Society* 39 (1956): 266–278.

12. M.D. Karkhanavala and F.A. Hummel, "Reactions in the System Li_2O-MgO-Al_2O_3-SiO_2: I, The Cordierite-Spodumene Join," *Journal of the American Ceramic Society* 36 (1953): 393–397.

13. S. Ray, "Solid Solutions in the Keatite Crystal Lattice," *Journal of the American Ceramic Society* 54 (1971): 213–215.

14. T.Y. Tien and F.A. Hummel, "Studies in Lithium Oxide Systems: XIII, $Li_2O.Al_2O_3.2SiO_2$-$Li_2O.Al_2O_3.2GeO_2$," *Journal of the American Ceramic Society* 47 (1964): 582–584.

15. F.A. Hummel, K. Auh, and G.H. Johnson, "Synthesis of the Chromium Analogue of Alpha Spodumene and its Solubility Relationships with Ureyite and Beta Spodumene," *Materials Research Bulletin* 5 (1970): 301–306.

16. F.A. Hummel, "Thermal Expansion Properties of Natural Lithia Minerals," *Foote Prints* 20 (1948): 3–11.

17. F.A. Hummel, "Thermal Expansion Properties of Some Synthetic Lithia Minerals," *Journal of the American Ceramic Society* 34 (1951): 235–239.

18. F.A. Hummel, "Significant Aspects of Certain Ternary Compounds and Solid Solutions," *Journal of the American Ceramic Society* 35 (1952): 64–66.

19. E.J. Smoke, "Ceramic Compositions Having Negative Linear Thermal Expansion," *Journal of the American Ceramic Society* 34 (1951): 87–90, and his "Ceramic Compositions Having Negative Linear Thermal Expansion: Part II," *Ceramic Age* (July 1953).

20. R.P. White and G.R. Rigby, "The Thermal Expansion Properties of Compositions Containing Lithia, Alumina, and Silica," *Transactions of the British Ceramic Society* 53 (1954): 324–34.

21. I.A. Bulavin and E.I. Medvedovskaya, "A Study of the Synthesis of Lithium Aluminosilicates," *Izvestiya Akademi Nauk SSSR, Neorganicheskie Materialy* 5 (1969): 1435–38 (in Russian).

22. F.H. Gillery and E.A. Bush, "Thermal Contraction of Beta Eucryptite ($Li_2O.Al_2O_3.2SiO_2$) by X-Ray and Dilatometer Methods," *Journal of the American Ceramic Society* 42 (1959): 175–177.

23. E.A. Bush and F.A. Hummel, "High Temperature Mechanical Properties of Ceramic Materials: II, Beta Eucryptite," *Journal of the American Ceramic Society* 42 (1959): 388–391.

24. W. Ostertag, G.R. Fischer, and J.P. Williams, "Thermal Expansion of Synthetic Beta Spodumene and Beta Spodumene-Silica Solid Solutions," *Journal of the American Ceramic Society* 51 (1968): 651–654.

25. J.H. Fishwick, "Treatment of Spodumene," U.S. Patent 3,394,988 (July 30, 1968).

26. C. Doelter, *Handbuch der Mineralchemie* 2 (1917): 193–204.

27. K. Endell and R. Rieche, *Zeitschrift fur Anorganische Chemie* 74 (1912): 33.

28. R. Ballo and E. Dittler, *Zeitschrift fur Anorganische Chemie* 76 (1912): 36.

29. F. Meissner, "Fusion and Transition Phenomena of Spodumene," *Zeitschrift fur Anorganische und Allgemine Chemie* 110 (1920): 187–95.

30. Roy, Roy, and Osborn, "Compositional and Stability Relationships."

31. G.G. Urazov, V.E. Plyushchev, and I.V. Shakhno, "The Monotropic Transformation of Spodumene," *Doklady Akademii Nauk SSSR* 113 (1957): 361-363.

32. A. Gabriel, M. Slavin, and H. Carl, *Econom. Geol.* 37 (1942) 116.

33. Endell and Riecke, *Zeitschrift.*

34. R. Barany and L.H. Adami, "Heats of Formation of Lithium Sulfate and Five Potassium and Lithium-Aluminum Silicates," U.S. Dept. of the Interior, RI Bur. of Mines Dept. 6873 (1966).

35. M.N. Sobolev, V.V. Lotov, and P.I. Asoskov, *Redkie Metally* 3 (1932): 47. F. Fraas and O. Ralston, U.S. Dept. Int. Bureau of Mines Rep. Invest., (3336), (1937). R. Hader, R. Nielsen, and M. Herre, *Industrial and Engineering Chemistry* (1951): 2636.

36. E.J. Smoke, "Inorganic Dielectrics Research," *Rutgers University Engineering Research Bulletin,* no. 50 (1970): 82.

4
Lithium in Glasses

The first recorded investigations on the use of lithium in glasses were performed in 1882 by Schott, who studied melts containing Li_2O, Na_2O, and SiO_2.[1] Although he did not succeed in his endeavor to produce new optical glasses, he did demonstrate the powerful fluxing action of lithia. In the United States, the first major application of lithium was in 1923 when D'Adrian patented the use of lepidolite (a lithium mica) as a fluxing constituent in ground coat and cover coat enamels.[2]

Fluxing Properties of Lithia

Lithium minerals and chemicals are more expensive than their sodium and potassium counterparts, and so, to obtain widespread application, they must demonstrate superior properties that make their use economical. The principal attribute of lithia in glass is its extremely powerful fluxing effect. This is attributed to the small size of the lithium ion and its high field strength. The ionic radius and ionic potential of the three common alkali metals are compared in Table 4.1.

Waterton and Turner studied the replacement of either Na_2O or K_2O by Li_2O in alkali-lime-silica glasses and found that lithia was the most effective alkali in reducing the viscosity.[3] They produced a hard frit at 500°C with a glass containing 15% Li_2O, whereas the corresponding Na_2O glass required 700°C and the glass containing 15% K_2O had to be melted at 850°C.

Table 4.1
A Comparison of the Ionic Potential of Li, Na, and K

Cation	*Valence*	*Ionic Radius (Å)*	*Ionic Potential*
Li^+	1	0.60	1.67
Na^+	1	0.95	1.05
K^+	1	1.33	0.75

The effectiveness of lithia as a flux in comparison with soda or potassia was again demonstrated by Dingwall and Moore, who measured the temperature of a soda-lime-silica glass corresponding to a viscosity of 10^{12} poises.[4] They found that the base glass gave 566°C, whereas the lithia, potassia, and soda glasses gave 500°C, 533°C, and 544°C, respectively.

Rauch et al. showed that the replacement of 1.0 weight percent Na_2O by 0.48 weight percent Li_2O caused reductions of 8.4% and 10.5% in melting time and 23.0% and 18.2% in fining time of zinc alabaster and opal glasses, respectively.[5]

These are some examples of the studies of lithia in glasses that were performed until 1960. Since that time, if the number of investigations reported in the ceramic literature and abstracts is a guide, there has been a resurgence in the use of lithia-containing raw materials in many different types of glasses.

Effect of Lithia on Glass Properties

Density. Lithium forms contracted or condensed glasses, which have densities higher than might be expected from calculation.[6] In fact, both spodumene and eucryptite form glasses that possess higher densities than the crystalline form of the minerals.

This tightening effect has found practical use in the ability of lithia to increase the surface hardening of glasses and glazes. For example, Koch et al. showed that the replacement of PbO, Na_2O, and K_2O by Li_2O or BeO increased surface hardness of a whiteware glaze by as much as 20%.[7] This increase in the surface hardness of glasses and glazes is related to the tightening effect of the oxygen ion in the presence of the lithium ion. Fajans and Kreidl showed that there is a progressive tightening, as indicated by lower molar refractions of O^{2-} in the series 2 Ba^{2+}, 2 Sr^{2+}, 4 Na^+, 2 Ca^{2+}, 4 Li^+, Al^{3+} - Na^+, Al^{3+} - Li^+.[8] Thus, the substitution of lithium for sodium, for example, results in the lower molar refraction of oxygen in a silicate glass.

Thermal Expansion. Many investigators have studied the effect of lithia on the thermal expansion of glasses. The results have been controversial. The thermal expansion factors published in 1894 by Winkelmann and Schott were criticized for attributing too low a value for lithia. Waterton and Turner published the following factors for the three common alkalies: Li_2O 4.9 $\times$ 10^{-7}, Na_2O 4.18 $\times$ 10^{-7}, and K_2O 3.4 $\times$ 10^{-7}.[9] These data indicate that, on a molar basis, lithia will reduce thermal expansion.

Navias summarized the results of early studies on thermal expansion and also concluded that the results were in error.[10] He proposed that, on a weight basis, lithia increases thermal expansion at least as much as soda.

Sun and Silverman reported that four different studies showed Li_2O:Na_2O expansion factor ratios of 1.16, 1.17, 1.33, and 1.63.[11] (For more information on the use of lithia for making low-expansion glasses, see p. 65.)

Surface Tension. Dietzel studied the effect of lithia on surface tension and concluded that it increased this property whether substituted for other alkali on a weight or a molar basis.[12]

Williams and Simpson studied the effect of lithium carbonate additions to silicate melts on their surface tensions using a maximum-bubble pressure method.[13] By molar replacement, Li_2O was added in 0.001, 0.01, 0.05, and 0.1 molar quantities to a soda-silica glass, a soda-lime-silica glass, a borosilicate glass, ground-coat and cover-coat enamels, and a whiteware glaze. The change in surface tension was found to be nonlinear, depending upon the composition of the base melt and the temperature, as well as the concentration of lithia. Lithia increased the surface tension of the glasses and enamels but decreased the surface tension of the glaze.

The similarity between lithium oxide and the alkaline earth oxides has been mentioned before. This similarity is again revealed by the effect of lithium on the surface tension of glasses. The alkali metals sodium and potassium tend to reduce surface tension, whereas, like alkaline earths, lithium raises it.

Shartsis and Capps demonstrated that lithium, sodium, and potassium all raise the surface tension of B_2O_3, each yielding an S-shaped curve.[14] These curves cross each other so that the order observed for surface tension in the low alkali range is $K > Na > Li$, whereas that in the high alkali range is $Li > Na > K$.

Electrical Properties. Most glasses have a high insulating value and a low dielectric loss. The insulating resistance of all glasses decreases as the temperature is increased until, in the fluid state, they become poor conductors.

Several investigators have demonstrated that the dielectric losses in glasses are mainly a result of alkaline ion mobility.[15] The alkali ions, loosely bound in the structure, oscillate more readily than the network under the influence of an applied potential and, therefore, absorb more energy.

Thurnauer and Badger showed that additions of lithia to a base glass (17.4 Na_2O, 10.1 CaO, 72.5% SiO_2) decreased the power factor; similar additions of soda caused an increase.[16] The additions were made on a molecular basis so that each glass contained an equal number of alkali ions.

Dale, Pegg, and Stanworth synthesized alumino-borosilicate glasses containing up to 15 weight percent Li_2O.[17] These glasses melted rapidly at 1200°C and had resistivities lower than ordinary soda-lime-silica glasses.

Lithia currently is being used in an increasing number of glasses to aid in melting and to improve physicochemical properties; however, there has not yet been an extensive study of the influence of lithium in electrical melting or of its effect on electrical boosting.

Chemical Durability. There have been no thorough investigations of the effect of lithia on the chemical durability of various glasses. A paper presented by the Owens-Illinois Glass Company reported that the weight substitution of lithia for soda in soda-lime glasses produced a moderate decrease in the resistance to attack by distilled water.[18] Simpson states that lithia offers the greatest durability of any of the alkali constituents of glass.[19] The resistance to attack by moisture is greatest with lithia because of the insolubility of its salts. The reason that lithia improves acid resistance is due to the fact that its powerful fluxing action can be used to keep the total alkali content to a minimum. Although the resistance of lithia to alkalies is not good, it is better than that of the other alkalies. Studies by this author, which have been mainly in the soda-lime-silica system, have demonstrated that the effect that lithia has on chemical durability depends on the manner in which it is introduced to the glass. Generally, weight substitutions for other alkalies leave the chemical durability of the base glass unchanged, while molecular substitutions improve the durability.

Finally, even a cursory examination of the ceramic literature reveals the presence of small amounts of lithia in a great number of glasses, glazes, and enamels for which high chemical durability is claimed.

Volatility. As for chemical durability, there have been no extensive investigations reported on the comparison of the volatilization of lithium with that of the other alkalies from commercial glasses. Preston and Turner studied the volatilization of Li_2O at 1300°C, 1350°C, and 1400°C from lithium oxide-silica glasses.[20] When compared on an equimolar basis (about 21 mole % R_2O), the initial rate of loss of alkali in mg/cm^2/20 hours was reported as 12.6, 5.1, and 3.0 for K_2O, Na_2O, and Li_2O glasses, respectively.

Soda-Lime Glasses

Over 90% of all glass produced is soda-lime glass, consisting of approximately 72% SiO_2, 15% Na_2O, 10% CaO and MgO, 2% Al_2O_3, and 1% miscellaneous oxides. This composition is used for making both container and flat glass.

One of the first extensive investigations of the effect of lithia in soda-lime glasses was performed by the Owens-Illinois Glass Company.[21] Lithia was substituted for soda in amounts of 2, 4, and 6 weight percent in two base glasses—one containing 16% Na_2O, 10% CaO.MgO, and 74% SiO_2 and the other containing 14% Na_2O, 12% CaO.MgO, and 74% SiO_2. The substitution of Li_2O for Na_2O in the 12% dolomite-lime glass caused a gradual reduction in liquidus temperature for the 2% and 4% substitutions; the liquidus temperature of the glass containing 6% Li_2O was identical to that of the base glass. This change in liquidus temperature is in a direction favorable for lithia since the viscosity was also reduced by lithia substitutions. The working range would, therefore, be virtually unchanged. No appreciable change in liquidus temperature was found when lithia was substituted for soda in the 10% dolomite glasses. The thermal expansion coefficient was substantially increased on substituting Li_2O for Na_2O. It should be noted, however, that the substitutions were made on a weight rather than a molecular basis, thus increasing the total number of alkali ions in the glass. This increase in the total number of alkali ions was also presumably a cause of the moderate decrease found in the resistance to attack by distilled water in both series on substituting lithia for soda.

Lester investigated the effects of lithia, substituted for soda on a weight basis, on the working characteristics of a container glass and found a substantial reduction in the low temperature viscosity.[22]

This author became interested in using spodumene for chemical boosting to solve one of the most important problems facing the container glass industry—that of increasing the productivity of glass melting furnaces without deteriorating the quality of the glass and without curtailing the campaign of the furnace.[23] This problem is normally solved by increasing the volume and area of the furnaces or by intensifying the melting by increasing the melting temperature. As the author points out, higher temperature melting will increase the vapor pressure of the batch materials, which will aggravate any existing air pollution problems and will probably reduce the furnace campaign because of accelerated refractory attack. In the experimental work, nine glasses were studied—a base glass (1) and eight lithia-substituted compositions (2–9). (See Table 4.2.) Four of the glasses (2–5) contained 0.1% to 0.4% lithia substituted for other alkalies on a weight basis, so that the total weight percent of alkali remained unchanged. The other four lithia-containing glasses (6–9) were formulated so that the total molecular percent of alkalies in the glasses remained unchanged. In the latter case, because of the lower atomic weight of lithia, the total weight percent of alkali decreased on increasing the percent of lithia. The calculated compositions of all nine glasses are given in Table 4.2. The glasses were melted in 500 ml platinum crucibles at 2650°F and fined at 2550°F. All glasses were fritted three times in water to ensure good homogeneity.

Table 4.2
Calculated Oxide Contents of Nine Experimental Container Glasses

Oxide	*1*	*2*	*3*	*4*	*5*	*6*	*7*	*8*	*9*
(%) SiO_2	72.97	72.96	72.95	72.94	72.93	73.06	73.14	73.23	73.31
Al_2O_3	1.700	1.700	1.700	1.700	1.700	1.700	1.700	1.705	1.709
Li_2O	0.000	0.100	0.200	0.300	0.400	0.100	0.200	0.300	0.400
Na_2O	13.90	13.86	13.83	13.79	13.75	13.75	13.60	13.45	13.30
K_2O	0.325	0.260	0.195	0.130	0.0656	0.261	0.196	0.131	0.0660
Fe_2O_3	0.0387	0.0514	0.0638	0.0765	0.0890	0.0514	0.0639	0.0768	0.0895
CaO	9.511	9.506	9.501	9.496	9.491	9.519	9.526	9.534	9.541
MgO	0.805	0.808	0.812	0.816	0.819	0.809	0.814	0.819	0.824
BaO	0.500	0.500	0.500	0.500	0.500	0.500	0.500	0.500	0.500
SO_3	0.230	0.230	0.230	0.230	0.230	0.230	0.230	0.230	0.230
As_2O_3	0.020	0.020	0.020	0.020	0.020	0.020	0.020	0.020	0.020

The substitution of lithia for other alkalies resulted in a decrease in the low temperature viscosity points. These are shown in Figure 4.1 and Table 4.3 along with other physicochemical properties. This decrease was, as expected, more marked for weight substitutions than for the molecular substitutions. On the average, each 0.1% of Li_2O decreased the low temperature viscosity points

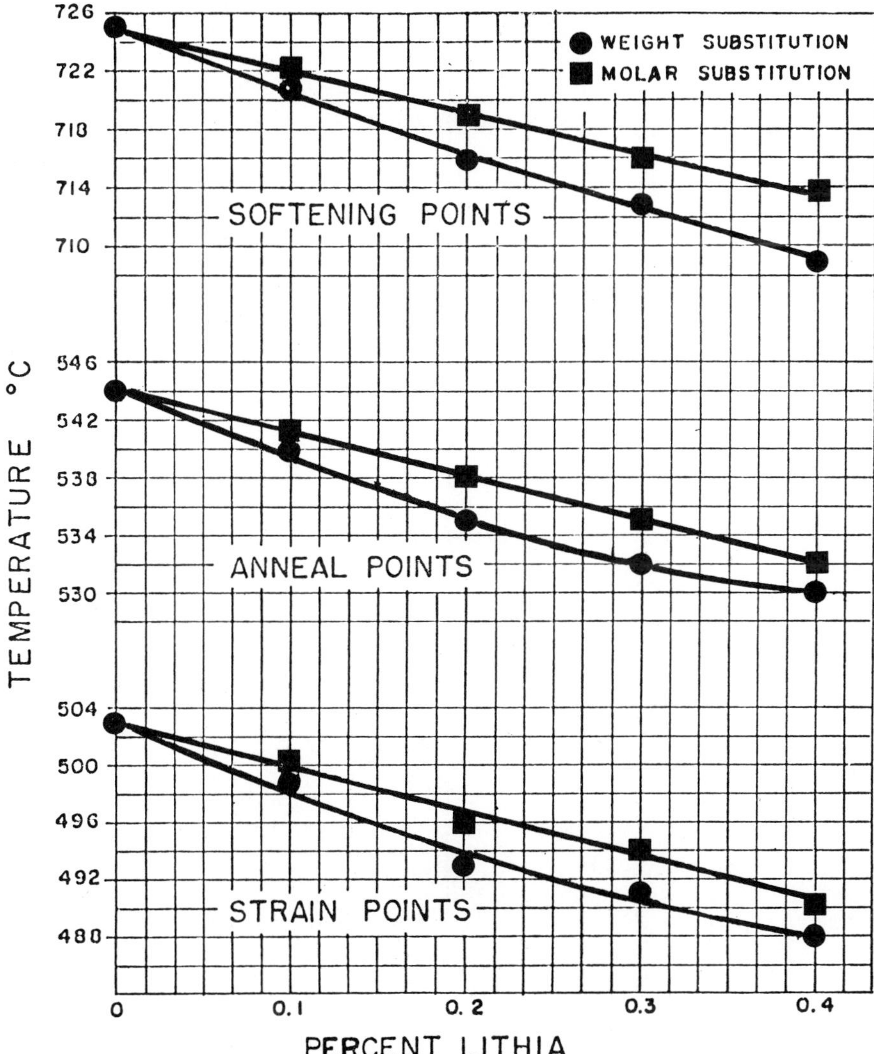

Figure 4.1. Effect of Lithia on Low Temperature Viscosity of Experimental Container Glasses

Table 4.3

Low Temperature Viscosities, Chemical Durabilities, Densities, Refractive Indexes, and Thermal Expansions of the Nine Studied Glasses

Glass Composition	*Strain Point (°C)*	*Anneal Point (°C)*	*Softening Point (°C)*	*Density (g/cc)*	*Refractive Index (Sod. D)*	*Thermal Expansion Room Temperature to 300°C × 10^{-7} in./°C*	*Chemical Durability (mls 1/50 N H_2SO_4)*
1. Base Glass	503	544	725	2.498	1.514	86.5	7.10
2. 0.1% Li_2O-wt	499	540	721	2.499	1.515	82.8	7.00
3. 0.2% Li_2O-wt	493	535	716	2.498	1.516	82.6	7.14
4. 0.3% Li_2O-wt	491	532	713	2.500	1.517	82.0	7.07
5. 0.4% Li_2O-wt	488	530	709	2.498	1.517	76.8	7.06
6. 0.1% Li_2O-mole	500	541	722	2.491	1.515	83.8	6.97
7. 0.2% Li_2O-mole	496	538	719	2.497	1.516	87.3	6.42
8. 0.3% Li_2O-mole	494	535	716	2.493	1.516	86.9	6.64
9. 0.4% Li_2O-mole	490	532	714	2.490	1.516	81.8	6.34

by 3.75°C and 3°C when added on a weight and molecular basis, respectively. There was virtually no change in the density or refractive index on substituting lithia in these glasses. The thermal expansion data were somewhat variable and were not sufficient to draw any definite conclusions. However, there appeared to be a slight reduction in this property with lithia additions. Chemical durability depends on the manner in which the lithia was introduced into the glass. Substitutions made on a molecular basis increased the durability, but weight substitutions left it unchanged. The effect of lithia on the high temperature viscosities is shown in Table 4.4. With the exception of glass 6, which appeared to be an anomaly, lithia reduced the high temperature viscosities of the glasses studied. The average reduction in temperature for each 0.1% Li_2O was 7.7°F and 3.5°F for the weight and molecular substitutions, respectively.

Realizing that the viscosity of a glass has a major influence on its melting and fining rate, this author performed melting rate and fining rate studies on the lithia-containing glasses and also measured the liquidus points. The melting rate studies were carried out using a gradient method on glass samples containing 0.17% Li_2O added on a weight basis. The measurements indicated that spodumene reduced the melting temperature by about 20°C or 36°F. Fining rate studies were performed by melting a glass containing 0.2% Li_2O, added on a weight basis, at 2650°F in platinum for 55–180 minutes. The number of seeds per ounce (N) present in cast and annealed disks was determined for several melting times (t). The results, which indicate that spodumene promoted faster fining, are given in Figure 4.2 as a regression of N on t.

The first commercial trials using lithia in container glass were made by Thatcher Glass Manufacturing Company. The results were reported by T.M. Mike.[24] After two brief trials, in which the addition of lithia as spodumene was observed to result in faster melting and increased productivity, a third, more extensive, trial was performed. This trial covered a period of about eighteen months until the furnace was shut down for repair. At 0.14% Li_2O, Mike concluded that the furnace was melting an extra ten tons per day because of the addition of lithia. The normal pull from this tank was 160 tons per day. When new, the pull from the tank was 3.6 square feet per ton. At the end of the campaign, the furnace was still able to pull 3.7 square feet per ton in spite of the fact that a major portion of the checkers had been mined out due to plugging.

Mike concluded that lithia additions to a container glass resulted in increased tonnage as well as lower melting temperatures and a prolonged furnace life. It was also believed that, because of lower melting temperatures, stack emission could be reduced for air pollution control.

The main purpose of adding spodumene in the commercial tank trials

Table 4.4
Effect of Lithia on High-Temperature Viscosity of Nine Experimental Container Glasses

	Temperature °F								
Log 10 Viscosity Poise	*Base Glass* 1	*0.1%* Li_2O *wt* 2	*0.2%* Li_2O *wt* 3	*0.3%* Li_2O *wt* 4	*0.4%* Li_2O *wt* 5	*0.1%* Li_2O *mole* 6	*0.2%* Li_2O *mole* 7	*0.3%* Li_2O *mole* 8	*0.4%* Li_2O *mole* 9
2.00	2652	2645	2638	2630	2620	2660	2645	2635	2640
2.20	2544	2537	2530	2522	2512	2551	2536	2524	2528
2.40	2445	2437	2430	2421	2412	2450	2435	2424	2428
2.60	2350	2344	2337	2327	2320	2356	2343	2331	2335
2.80	2265	2260	2253	2244	2235	2270	2259	2249	2252
3.00	2186	2180	2173	2165	2155	2190	2180	2170	2173
3.20	2114	2108	2101	2093	2084	2119	2108	2097	2100
3.40	2045	2040	2034	2025	2017	2050	2040	2029	2032

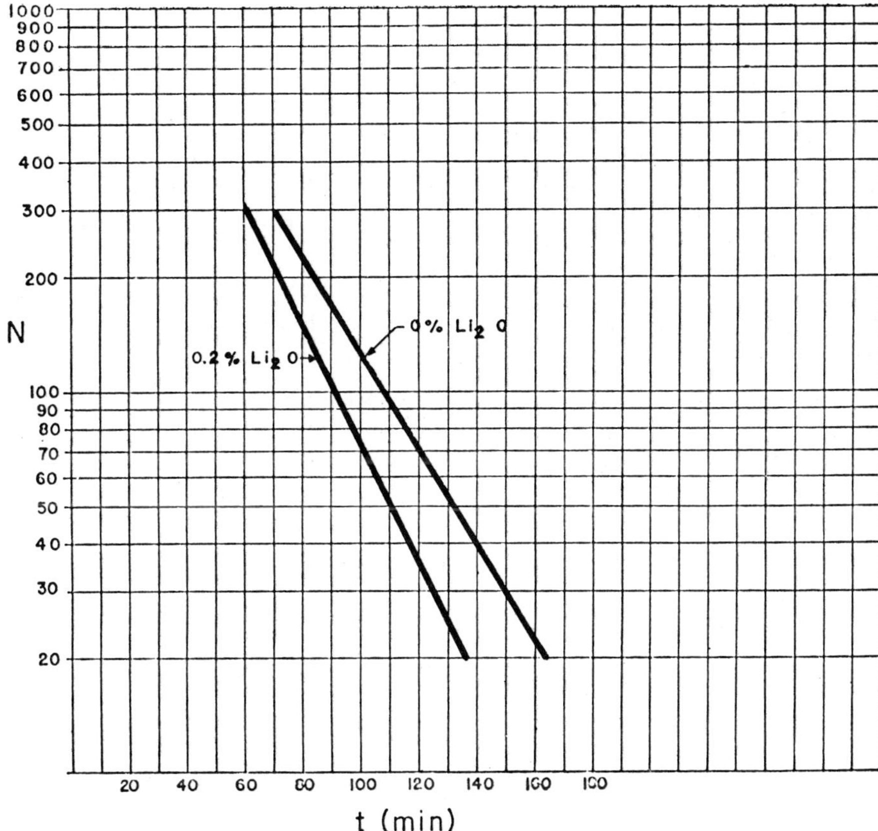

Figure 4.2. Fining Rate of a Glass Containing 0.2 Weight Percent Li_2O as a Regression of *N* (Number of Seeds per Ounce) on *t* (Melting Times in Minutes)

described above was to increase the pull, permitting the furnace campaign to be extended by as much as several months. The author believes that spodumene can also be used economically in the first half of a campaign to reduce melting temperatures and stack emissions.

The use of lithium minerals to reduce or eliminate air pollution is a relatively new concept, which is growing in importance as regulations become more stringent. This author has recently studied the replacement of fluorspar by spodumene in container glass batches and concluded that a 1:1 replacement leaves the low temperature viscosity points and batch melting rate unchanged. Spodumene is already being used commercially in making foam glass, where it is able to successfully replace B_2O_3 and CaF_2 and thus eliminate an air pollution problem.

Fiber Glasses

Although many fiber glass compositions are known, commercial formulations comprise essentially two types: insulation grade (high alkali) and textile grade (low alkali). The insulation grade, or T-glass, is produced from relatively low-cost raw materials and is used for structural insulation for pipes and ducts. The textile grade, or E-glass, is more expensive and is used to reinforce plastic and tire cord.

The powerful fluxing action of Li_2O and B_2O_3 on silica was demonstrated by Sastry and Hummel in their work on lithium oxide systems.[25] This fluxing action is currently being used to advantage in commercial fiber glass production for facilitating melting and for eliminating fluorspar from the compositions and thus avoiding air pollution problems. Because fiber glasses can tolerate some small amount of iron, spodumene is the preferred source of lithia. Volf has summarized some fiber glass compositions containing lithia. These are shown in Table 4.5 [26]

Welsch studied several fiber glass compositions with the goals of providing low operating temperatures, high fiber production rates, a low volatilization rate, a low liquidus point, and a slow devitrification rate.[27] He noted that small amounts of lithia tend to improve glass durability. He reported the composition shown in Table 4.6, which has a liquidus temperature of 1685°F and a temperature of 1870°F at 1000 poises.

Glass fibers suitable for reinforcing cement products have been made and patented by Pilkington Brothers Ltd.[28] These fibers contain lithia and are

Table 4.5
Several Fiber Glasses Containing Li_2O

	Percent				
Oxide	*1*	*2*	*3*	*4*	*5*
SiO_2	58	60	70	58	79.4
B_2O_3	5	–	22	–	15.1
Al_2O_3	12	15	2	9	–
CaO	11.5	6	–	–	1.6
MgO	12	4	–	30	–
Na_2O	–	–	1	–	1.2
K_2O	–	–	2	–	–
Li_2O	0.5	5	3	2	1.1
Sb_2O_3	–	–	–	–	1.7
CaF_2	1.0	10	–	–	–

Source: M.B. Volf, *Technical Glasses* (London: Sir Isaac Pitman and Sons, 1961).

Table 4.6
Li_2O-Containing Fiber Glass

Oxide	*Percent*
SiO_2	60.8
Al_2O_3	3.3
CaO	7.7
MgO	3.7
Na_2O	14.5
B_2O_3	8.6
K_2O	1.0
Li_2O	0.5

Source: W.W. Welsch, "Glass Composition," U.S. Patent 2,877,124 (March 10, 1959).

reported to have improved resistance to alkali attack. A typical composition is: SiO_2 69, ZrO_2 9.5, Fe_2O_3 15.5, Li_2O 2, CaO 2, and TiO_2 2 weight percent; the fiberizing temperature is 1270°C and the liquidus temperature is 1110°C.

Bastian and Ottoson studied a large number of experimental glass compositions in an effort to make fibers with a high Young's modulus.[29] The basic concept of their invention is enhanced by the substitution of lithium oxide as a flux for silica to replace sodium or potassium oxides. When used in the suggested range, lithia can eliminate the detrimental effect of sodium. The authors claim unleached glass fibers having a diameter of less than 0.001 inch with a Young's modulus of at least 15 million, consisting of R oxide 12–24, lithium oxide 2–6, beryllium oxide 5–12, silica balance, where R is Ca or Mg or a mixture.

Thomas has patented the following lithia-containing high tensile strength magnesium aluminum silicate glass compositions: SiO_2 50.0–64.0, Li_2O 0.1–2.5, Al_2O_3 18.0–30.0, MgO 11.0–23.0, B_2O_3 0.0–4.9, Sb_2O_3 0.0–1.0, and Fe_2O_3 0.0–1.0.[30] These fibers have a tensile strength of from 541,000 to 663,000 psi.

A major breakthrough in the manufacture of fiber glass would be the development of a suitable composition that could be processed through bushings made of a material other than platinum. Suggestions that a high-lithia glass may be the basis of such a composition have been made, but no practical compositions have yet been reported.

Television Tube Glasses

According to Volf, the development of glasses for black and white television tubes ended with the lithium-barium glasses, upon which world production

Table 4.7
Composition Range of Television Tube Glasses

Oxide	*Percent*
SiO_2	65.3–69.1
Al_2O_3	3.2– 6.2
CaO	0– 0.6
MgO	0– 0.3
BaO	11.4–12
K_2O	6.6–7
Na_2O	7.2–8
Li_2O	0.6–0.7
Sb_2O_3	0–0.4
F	0–1

is presently based.[31] Lithium oxide is an important component of these glasses and is responsible for reducing the viscosity and improving the stability and electrical properties of these glasses. The range of compositions used is shown in Table 4.7. These lithium-barium glasses have a viscosity curve close to that of lead glasses. The presence of BaO in the glass prevents emanation of harmful radiation from the screen. It is generally accepted that lithium-barium glasses are able to protect human health because of their ability to absorb harmful radiation. Some actual lithium-containing compositions for television tube glasses are shown in Table 4.8.

Radiation Glasses

The small size and high field strength of the lithium ion has again been used to advantage in stabilizing glasses to solarization. Thus, White and Silverman found that lithium glasses exhibited no color change when subjected to mercury-arc radiation.[32]

Florence et al. measured the transmittance of infrared energy by lithium, sodium, potassium, and lead silicate glasses for wavelengths of 0.7 to 5 microns and concluded that lithium silicate compositions gave a greater transmittance than glasses in either the soda-silica or potash-silica systems.[33]

This author has studied the use of lithium for developing glasses capable of attenuating thermal neutrons. Lithium possesses a high cross-section for thermal neutrons, much higher than that of the other alkali metals.[34] Also, lithium does not give rise to highly energetic secondary gamma radiation during the process of neutron capture. This is a distinct advantage where the protection

Table 4.8
Some Lithia-Containing Television Tube Glasses and Their Properties

Oxide	*Corning, U.S.A.*	*Sovirel, France*	*Schott, Mainz*	*Phillips, Holland*	*Mullard, England*	*U.S.S.R.*	*Meiwa, Japan*	*GEC, England*
SiO_2	64.5	67.5	65.3	69.1	67.8	67.5	67.5	67–69
Al_2O_3	5.3	4.7	6.2	3.2	5.6	5.0	4.8	1.5–2.5
CaO	2.5	–	0.5	0.6	–	–	–	6.5–7.5
MgO	–	–	0.3	0.2	–	–	–	–
BaO	12.8	12	12.7	11.4	11.7	12.0	11.2	4.5–5.5
Na_2O	7.0	7.2	8.4	8.8	8.1	7.0	7.0	7–8
Li_2O	1.5	0.6	0.6	0.6	0.6	0.6	0.7	0.4
Sb_2O_3	–	0.4	–	–	–	0.4	–	–
F	–	–	–	–	–	0.9	–	1
$\alpha_{20-300} \times 10^7$	90	93.1	91.9	95.5	93.7	89.5	92.5	
Transformation temperature	490	450	450	468	475	512	442	
Deformation temperature	500	487	482	500	515	–	487	
Softening point	665	655.9	654.6	669.6	–	–	653	
Flow point	–	880	878	–	–	–	880	
tan $\delta \times 10^4$	–	21.9	26.0	22.7	25.3	–	22.3	
Modulus of elasticity	–	5804	5781	5639	5821	–	5424	
Density	2.7	2.64	2.59	2.56	2.62	2.58	2.60	

of personnel is concerned. Two glasses of the following compositions were studied: composition 1: SiO_2 20, Li_2O 20, B_2O_3 50, Al_2O_3 9.5, and CeO_2 0.5; composition 2: SiO_2 15, Li_2O 25, B_2O_3 59.5, and CeO_2 0.5. The glasses were cast into slabs, which were then polished and tested for neutron attenuation using center line fast neutron dose rates with an 8 inch thick polyethylene slab in position around the glasses. The results, which are given in terms of fast-neutron dose rate transmission fractions D/D_0 for various thicknesses, are shown in Figure 4.3. Note that D is transmitted dose rate for a particular glass

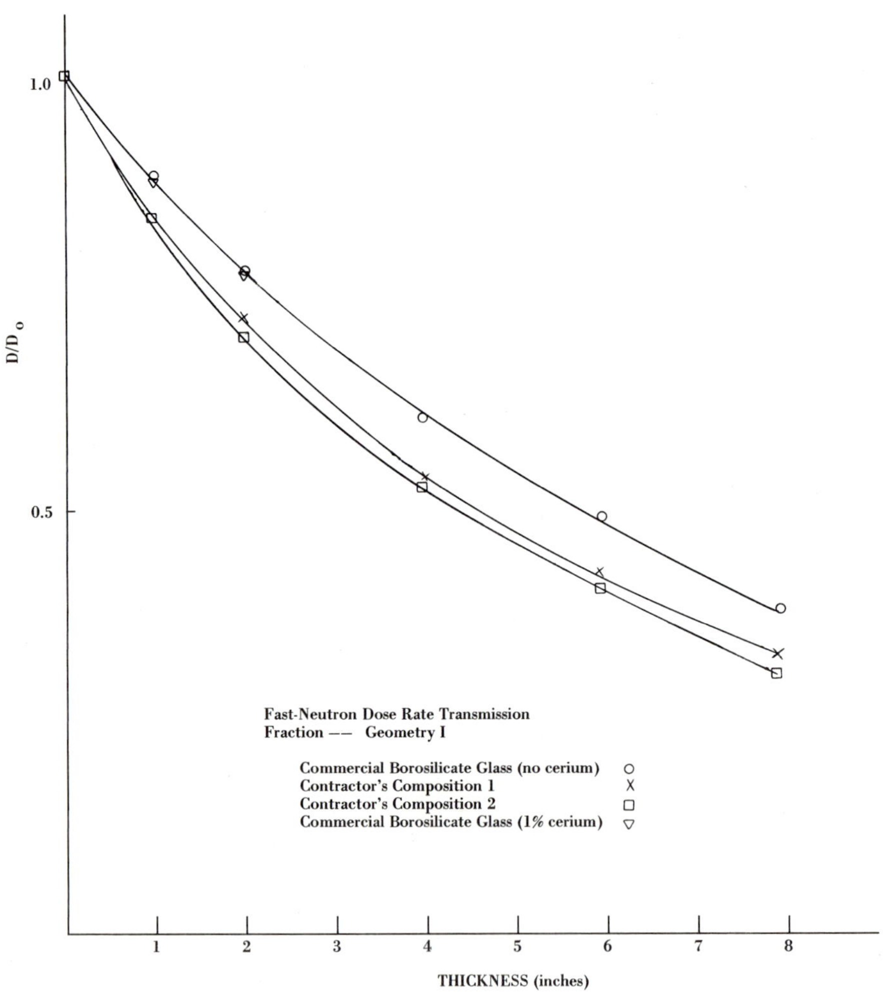

Figure 4.3. Fast Neutron Dose Rate Transmission for Lithia Glasses

thickness and D_0 is the dose rate for the condition of no glass sample in position.

For optimum attenuation of the radiation spectrum expected one to two miles from a tactical nuclear weapon explosion, the presence of high atomic weight elements is mandatory for the scattering of high energy neutrons. To meet these requirements, this author developed glasses in the system Li_2O-PbO-B_2O_3-SiO_2 at 10%, 15%, and 20% Li_2O. An extensive area of stable glasses was found in this system that should possess good radiation-absorbing properties. The three best compositions were PbO 10, 30, 50; Li_2O 15, 15, 10; B_2O_3 50, 40, 25; SiO_2 25, 15, 15; and CeO_2 (above 100%) 0.5, 0.5, and 0.5 weight percent.

Sugiura et al. developed glasses in the Li_2O-PbO-B_2O_3 system suitable for radiation shielding properties, but the chemical durability of the glasses was poor.[35]

More recently, Bobkova studied the region of glass formation in the system Li_2O-PbO-B_2O_3.[36] He made glasses containing B_2O_3 50–90, Li_2O 0–40, and PbO 5–40 mole percent and found that a glass containing B_2O_3 70, PbO 25, and Li_2O 5 mole percent had the best fusibility.

Low-expansion Glasses

Various compositions within the Li_2O-Al_2O_3-SiO_2 system can be melted to form low-expansion glasses. For example, Brackbill et al. reported the thermal expansion of glasses ranging in molecular ratios from 1:1:2 (Li_2O:Al_2O_3:SiO_2) to 1:1:10 to vary from 76.0 to 39.3 $\times 10^{-7}/°C$ in the range 30°C to 500°C.[37] The authors point out that such compositions may serve as a basis for developing high-fire glazes for use on mullite, alumina, zircon, and kyanite porcelains.

Hood patented a low-expansion lithia-containing glass suitable for making large telescope mirrors and other massive articles.[38] This glass contains 0.5–2.5% alkali oxide, of which lithia constitutes up to one third of the total alkali.

Most commercial low-expansion glasses that are used for heat-shock applications contain substantial amounts of boron. Dubrovo and Shmidt reported some glass compositions that contain little or no boron oxide and yet exhibit very low thermal expansions.[39] All of these compositions contain small amounts of lithia as shown by the compositions shown in Table 4.9.

Many years ago Taylor patented low-expansion lithia-boron glasses.[40] The compositions and thermal expansions of three such glasses are shown in Table 4.10.

Table 4.9
Compositions and Expansions of Some Low-Boron Glasses Containing Lithia

Oxide and Thermal Expansion	Percent					
	1	2	3	4	5	6
SiO_2	71	71	71	71	71	72
B_2O_3	–	–	–	3	2	3
Al_2O_3	10	10	9	8	9	9
CaO	4	4	6	8	6	5
CaF_2	7	7	5	2	4	4
MgO	6	6	6	6	6	4
Na_2O	–	1	2	–	–	1
Li_2O	2	1	1	2	2	2
$\alpha_{20-400} \times 10^{-7}/°C$	44.6	45.6	50.4	44.2	43.6	46.3

Source: S.K. Dubrovo and Yu. A. Shmidt, *Zhurnal Prikladnoi Khimii* 32 (1959): 742.

Table 4.10
Compositions and Expansions of Some High-Boron Glasses Containing Lithia

Oxide	Glass 1	Glass 2	Glass 3
SiO_2	71	75	70
B_2O_3	28	15	13
Li_2O	1	1	9
Al_2O_3	–	5	2
Na_2O	–	4	–
Sb_2O_3	–	–	6
$\alpha \times 10^{-7}$	29	40	56

Source: W.C. Taylor, "Glass," U.S. Patent 1,192,474 (July 25, 1916).

Hardwerk and McVay studied the thermal expansion of some glasses in the system Li_2O-CaO-SiO_2 and reported that, while a weight substitution of soda for lithia lowered the thermal expansion, a molecular substitution of soda for lithia increased it.[41] This again demonstrates that, on a molecular basis, lithia reduces the expansion of glasses because of the smaller total weight of alkali.

Electrode Glasses

Sokolov and Passinskij employed lithium and barium for manufacturing electrode glasses.[42] They used the glass compositions shown in Table 4.11.

Table 4.11
Lithia-Containing Electrode Glasses

Oxide	*Glass 1*	*Glass 2*
SiO_2	74	74
MgO	10	–
BaO	–	10
Li_2O	16	16

Source: Volf, *Technical Glasses.*

In the United States, electrode glasses were produced with the composition: SiO_2 67%, BaO 8%, and Li_2O 25%.[43]

According to Volf, the smaller the ionic radius of a univalent element in the glass, the greater its mobility. This enhances the electrical conductivity of the glass.[44] Because BaO increases electrical resistivity, glasses were sought that would contain little or no barium but would be stable at elevated temperatures. Perley reported two such glasses; these compositions are given in Table 4.12.[45]

Table 4.12
Perley Glasses

Oxide	*Glass 1*	*Glass 2*
SiO_2	63	63
La_2O_3	2	2
Cs_2O	2	2
Li_2O	25	25
CaO	8	–
BaO	–	8

Source: Ibid.

Sealing Glasses

Volf divides iron-sealing glasses into four groups, the first of which is called the lithium-containing glasses. These contain 10–38% PbO, 10–20% BaO, and a combination of sodium, potassium, and lithium oxides.[46] Several lithia glasses are shown in Table 4.13.

Meister and Schiveree studied seals for electron discharge lamps consisting of fused quartz envelopes and refractory leads. They recommended a petalite

Table 4.13
Iron-Sealing Glasses Containing Lithia

Oxide	*1*	*2*	*3*	*4*	*5*	*6*
SiO_2	45	45	45	40	38.7	40–48
PbO	25	17.5	10	30	30	32–38
BaO	10	17.5	25	20	20	–
Al_2O_3	–	–	–	–	–	1–8
Na_2O	5.4	5.4	5.4	2.5	7	2–5
K_2O	12.6	12.6	12.6	7	1.8	12–17
Li_2O	2	2	2	0.5	0.5	1–25
$\alpha_{200-300} \times 10^7$	123.3	121.7	128.2	106.6	107	–
$\tan \delta \times 10^4$	4.3	4.3	4.3	4.3	4.3	
ϵ	8.2	8.3	8.3	8.8	8.8	

Source: Ibid.

glass for making these seals, which eliminates the need for expensive graded seals.[47]

Low Dielectric Loss Glasses

According to Volf, the heat due to dielectric loss can increase the temperature of glass components used in circuits operating at high voltages and high frequencies, thus producing permanent deformation.[48] Thus, special low-dielectric loss glasses are needed for high-frequency circuits. Lithium is an important glass constituent because of its small ionic radius and strong electrostatic field. Although lithium is more mobile than sodium (and, therefore, gives rise to a greater loss angle), when lithium is molecularly substituted for soda, $\tan \delta$ diminishes because of the contraction effect. Several lithia-containing low-loss lead glasses are shown in Table 4.14.

Borosilicate Opal Glasses

Flannery and Lemoine patented an opal borosilicate glass exhibiting a very bright white appearance and possessing excellent resistance to detergents.[49] The opal phase is present as discontinuous, spherically shaped droplets. The compositional range claimed is: 0.5–2.5% Li_2O, 7–10% ZrO, 11–14% B_2O_3, and 71–76% SiO_2. These glasses are suitable for tableware and ovenware because they combine good thermal shock resistance with good chemical durability.

Table 4.14
Lithia-Containing Low-Loss Lead Glasses

	1	*2*	*3*	*4*	*5*	*6*	*7*
SiO_2	40.0	40.0	40.0	39.5	43.2	42.0	50.0
PbO	50.0	50.0	50.0	49.3	43.9	40.0	30.0
K_2O	6.3	5.6	6.65	6.2	8.6	11.5	12.9
Na_2O	2.7	2.4	2.85	2.0	3.5	5.0	5.6
Li_2O	1.0	2.0	0.5	1.0	0.8	1.5	1.5
Na_2SiF_6	–	–	–	2.0	–	–	–
$\tan \delta \times 10^4$	4.6	5.1	5.0	4.2	4.6	4.6	5.1
$\alpha \times 10^7$	103	104	103	102	–	128	122
Softening point	521°	506°	534°	505°	–	503°	521°

Source: Ibid.

Photochromic Glasses

Photochromic spectacle lenses were developed by Corning Glass Works to darken under ultraviolet light. They are made from an alkali aluminoborosilicate glass containing a silver halide, which is the light-sensitive agent. The addition of lithium carbonate controls the alkalinity of the glass and the solubility of the silver halides. Lithium carbonate is the preferred source of lithium for this application.

Chemical Strengthening

Glass is intrinsically very strong, but, because of surface imperfections, the actual tensile strength rarely exceeds 6,000–7,000 psi. There are several methods used for increasing the strength of glass surfaces. These provide compressive forces in the surface of the glass which have to be overcome before the glass will fail under tension. The principal methods are:

1. Cladding with a lower expansion material.
2. Thermal tempering.
3. Ion exchange.
4. Surface nucleated crystallization.

As long ago as 1891, Schott made high-strength, boiler-gage tubing by overlaying a low-expansion glass on a high-expansion glass. The greater contraction of the core on cooling placed the surface in compression and the core in tension. This method is currently being used to improve the strength of certain types of glazed dinnerware.

Until the last few years, physical or thermal tempering was the most important method for strengthening glass. This involves quenching a glass from just below its softening point, which again results in surface compression because the interior cools more slowly and continues to contract after the exterior has become rigid.

The last two methods, ion exchange and surface crystallization, are relatively new and frequently involve the use of lithium. The ion exchange method for chemically strengthening glasses can be carried out at a low temperature, below the annealing point, or at a high temperature, above the annealing point but below the softening point. In the former process, large ions, such as sodium or potassium, are "stuffed" into surface lattices occupied by a smaller ion such as lithium. This places the surface into compression. In the latter process, the glass is heated in a lithium-containing salt bath to effect a lithium replacement in the glass, which produces a lower expansion in the surface layers. Surface nucleated crystallization involves the devitrification of a lithia-alumina-silica glass to produce crystals with a low thermal expansion coefficient.

Hood and Stookey were able to increase the mechanical strength of glass by contacting a glass above its strain point but below its softening point with a lithium salt.[50] In this process, lithium ions migrate or diffuse into the glass in exchange for sodium or potassium ions. The weight of lithium as Li_2O introduced into the glass is about one-half of the weight of displaced Na_2O plus one-third of the weight of displaced K_2O. This results in a decrease in the total weight percent of alkali oxide in the surface layers, which reduces the thermal expansion.

Olcott and Stookey patented a glass body on which a thin, compressive semicrystalline layer was produced.[51] This method comprises heat treating a glass consisting of 52–65 parts of SiO_2, not less than 4 parts of Li_2O, and not more than 40 parts of Al_2O_3 at a temperature at which its viscosity is between 10^7 and 10^{10} poises. This results in the formation of a semicrystalline layer of beta eucryptite, which has a thermal expansion considerably lower than that of the glass.

Stookey and Sawchuck patented the manufacture of transparent, high-strength ceramic bodies by providing a crystalline surface layer.[52] Here, a glass consisting of 56–73 SiO_2, 4–7 TiO_2, 8.5–20 R_2O, and 11–24 mole percent Al_2O_3 was contacted with a lithium salt-containing material at a temperature between 800°C and 900°C.

Plumat patented a salt bath for supporting glass ribbon in the plastic state which consists of 13–18% potassium chloride, 7–13% lithium chloride, 20–30% sodium chloride, and 40–60% barium chloride.[53]

The chemical strengthening of a lithia-alumina-silica glass by ion exchange with a larger ion was patented by Chisholm et al.[54]

Finally, Shonebarger patented a different method for strengthening glass.[55] Rather than exchanging or replacing ions, molecules of LiOH or KOH were diffused into the surface layers of the glass. As a result, the alkali metal hydroxide molecules crowd the surface layer, setting up a compression zone. In contrast to the ion exchange method, this crowding procedure is not carried out above the strain point. Other compounds, such as lithium amide, $LiNH_2$, can also be used. Here, the amide is decomposed by water vapor to give LiOH at the site of migration.

Summary

Lithium can be used in many glass formulations for increasing production rates, lowering thermal expansion, improving chemical durability, and altering radiation properties. It is becoming widely used as a flux in replacement for boron and fluorine where air pollution problems exist. Generally, the reason lithium is selected for these applications is a result of its small ionic radius and high field strength in comparison with the other alkali metals.

Notes

1. H. Hovestadt, *Jena Glass and Its Scientific and Industrial Applications,* trans. J.D. and Alice Everett (New York: Macmillan Company, 1902), 6–14.

2. A.L.D. D'Adrian, "Use of Lepidolite in Enamels," U. S. Patent 1,443,813 (January 30, 1923).

3. S.C. Waterton and W.E.S. Turner, "Some Properties of Mixed Alkali-Lime-Silica Glasses Containing Lithia, Soda, Potash, and Rubidia," *Journal of the Society of Glass Technology* 18 (1934): 268–85T.

4. A.G.F. Dingwall and H. Moore, "Effects of Various Oxides on Viscosity of Glasses of the Soda-Lime-Silica Type," *Journal of the Society of Glass Technology* 37 (1953): 316–72T.

5. H.W. Rauch, C.H. Commons, Jr., and A. Silverman, "Effect of Partial Replacements of Soda by Lithia in Opal and Alabaster Glass," *Glass Industry* (May 1960).

6. F.A. Hummel, "Thermal Expansion Properties of Natural Lithia Minerals," *Foote Prints* 20 (1948): 3–11.

7. W. Koch, C. Harmon, and L. O'Bannon, "Some Physical and Chemical

Properties of Experimental Glazes for Vitrified Institutional Whiteware," *Journal of the American Ceramic Society* 33 (1950): 1–8.

8. K. Fajans and N.J. Kreidl, "Stability of Lead Glasses and Polarization of Ions," *Journal of the American Ceramic Society* 31 (1948): 105–114.

9. Waterton and Turner, "Some Properties."

10. L. Navias, "Lithium Oxide as a Constituent of Glasses: Its Effect on Thermal Expansion," *Journal of the American Ceramic Society* 18 (1935): 206–10.

11. K.H. Sun and A. Silverman, "Additive Factors for Calculating the Coefficient of Thermal Expansion of Glass from Its Composition," *Glass Industry* 22 (1941): 114–15, 125.

12. A. Dietzel, "Practical Importance and Calculation of Surface Tension of Glasses, Glazes, and Enamels," *Sprechsaal* 75 (1942): 82–85.

13. R.M. Williams and H.E. Simpson, "A Note on the Effect of Lithium Oxide upon the Surface Tension of Several Silicate Melts," *Journal of the American Ceramic Society* 34 (1951): 280–283.

14. L. Shartsis and W. Capps, "Surface Tension of Molten Alkali Borates," *Journal of the American Ceramic Society* 35 (1952): 169–172.

15. J.M. Stevels, "Relation Between Dielectric Losses and Composition of Glass: I, Theoretical Introduction," *Journal of the Society of Glass Technology* 34 (1950): 80–86T, "II, Comparison of Theoretical Considerations with Experimental Results," ibid., 86–93T, "III, Behavior of Dielectric Losses in a Number of Special Cases," ibid., 93–96T, "IV, General Picture of Dielectric Losses in More Complicated Glasses as a Function of Their Composition," ibid., 96–100T; W.A. Weyl, "Dielectric Properties of Glass and Their Structural Interpretation," *Journal of the Society of Glass Technology* 33 (1949): 220–38T.

16. H. Thurnauer and A.E. Badger, "Dielectric Loss of Glass at High Frequencies," *Journal of the American Ceramic Society* 23 (1940): 9–12.

17. A.E. Dale, E.F. Pegg, and J.E. Stanworth, "Electrical Properties of Some Lithia-containing Glasses," *Journal of the Society of Glass Technology* 35 (1951): 136–45T.

18. Owens-Illinois Glass Company, "Effect of K_2O and Li_2O on Properties of Soda-Dolomite-Lime-Silica Glasses," *Journal of the American Ceramic Society* 33 (1950): 181–186.

19. H.E. Simpson, "Lithium Development and Expansion," *Glass Industry* 36 (1955): 17–22, 46–47.

20. E. Preston and W.E.S. Turner, "The Volatilization of Lithium Oxide at High Temperatures from Lithium Oxide–Silica Glasses," *Journal of the Society of Glass Technology* 18 (1934): 143–168.

21. Owens-Illinois, "Effect of K_2O and Li_2O."

22. W.R. Lester, "The Influence of Minor Additions of Lithia on The Working Characteristics of a Container-Type Glass," *Central Glass and Ceramic Research Institute Bulletin* 7 (1960): 34–37.

23. J.H. Fishwick and R.W. Talley, "Laboratory and Tank Data on the Effects of Lithia in Container Glass," *Foote Prints* 38 (1970): 13–17; J.H. Fishwick, "Lithia Reduces Viscosity, Melting, and Firing Times," *Glass Industry* (September–October 1972).

24. T.M. Mike, "Chemical Boosting of a Glass Melting Furnace Using Spodumene," paper presented at the Glass Problems Conference, Urbana, Illinois, (November 1971).

25. B.S.R. Sastry and F.A. Hummel, "Studies in Lithium Oxide Systems: VII, Li_2O-B_2O_3-SiO_2," *Journal of the American Ceramic Society* 43 (1960): 22–33.

26. M.B. Volf, *Technical Glasses* (London: Sir Isaac Pitman and Sons, Ltd., 1961).

27. W.W. Welsch, "Glass Composition," U.S. Patent 2,877,124 (March 10, 1959).

28. P.S. Irlam and B. Yale, "Glass Compositions for Fibermaking," South African Patent 7100,497.

29. R.R. Bastian and A.C. Ottoson, U.S. Patent 2,978, 341 (April 4, 1961).

30. G.L. Thomas, "High Tensile Strength Magnesium Aluminum Silicate Glass Compositions," U.S. Patent 3,189,471 (May 11, 1962).

31. Volf, *Technical Glasses.*

32. J.F. White and W.B. Silverman, "Some Studies on the Solarization of Glass," *Journal of the American Ceramic Society* 33 (1950): 252–257.

33. J.M. Florence, F.W. Glaze, C.H. Hahner, and R. Stair, "Transmittance of Near-Infrared Energy by Binary Glasses," *Journal of the American Ceramic Society* 31 (1948): 328–331.

34. J.H. Fishwick, "Manufacture and Evaluation of Two Experimental Neutron Attenuation Glasses," Report 1–W94–3 (January 25, 1964), and his "Radiation-Absorbing Glass," Report R–1778 (October 1965), both in the possession of the author.

35. T. Sugiura, K. Murakami, H. Tanaka, and H. Akimoto, "Radiation Shielding Materials: II, Radiation Shielding Properties of Glasses in the System Li_2O-PbO-B_2O_3," *Yogyo Kyokai Shi* 72 (1964): 71–75.

36. N.M. Bobkova, "Glass Formation in the Lithium Oxide-Lead Oxide-Boric Oxide System," *Stekloobraznye Sistemy: Novye Stekla na ikh Osnove* (1971): 95–97.

37. C.E. Brackbill, H.A. McKinstry, and F.A. Hummel, "Thermal Expansion of Some Glasses in the System Li_2O-Al_2O_3-SiO_2," *Journal of the American Ceramic Society* 34 (1951): 107–109.

38. H.P. Hood, "Low Expansion Glass Containing Lithia," Britain Patent 446,733 (1936).

39. S.K. Dubrovo and Yu. A. Shmidt, *Zhurnal Prikladnoi Khimii* 32 (1959): 742.

40. W.C. Taylor, "Glass," U.S. Patent 1,192,474 (July 25, 1916).

41. J.H. Handwerk and T.N. McVay, "Thermal Expansion of Some Glasses in the System Li_2O-CaO-SiO_2," U.S. Atomic Energy Commission Report No. ORO–216 (July 16, 1954).

42. S.M. Sokolov and A.H. Passinskij, *Zeitschrift fur Physikalische Chemie* 160 (1932): 366.

43. G.A. Perley, *Analytical Chemistry* 21 (1949): 391.

44. Volf, *Technical Glasses.*

45. Perley, *Analytical Chemistry* 21.

46. Volf, *Technical Glasses.*

47. G. Meister and J.E. Schiveree, "Quartz Metal Seal," U.S. Patent 2,675,497 (April 13, 1954).

48. Volf, *Technical Glasses.*

49. J.E. Flannery and J.G. Lemoine, "Borosilicate Opal Glasses," U.S. Patent 3,723,144 (March 27, 1973).

50. H.P. Hood and S.D. Stookey, "Method of Making a Glass Article of High Mechanical Strength and Article Made Thereby," U.S. Patent 2,779,136 (January 29, 1957).

51. J.S. Olcott and S.D. Stookey, "Glass Body Having a Semicrystalline Surface Layer and Method of Making It," U.S. Patent 3,253,975 (May 31, 1966).

52. S.D. Stookey and L.G. Sawchuck, "Transparent Devitrified Strengthened Glass Article and Method of Making It," U.S. Patent 3,282,770 (November 1, 1966).

53. E. Plumat, "Protective Salt Bath for a Glass Ribbon," U.S. Patent 3,295,945 (January 3, 1967).

54. R.S. Chisholm, G.E. Sleighter, and F. Ernsberger, "Method of Increasing Stain Resistance of a Glass Body Subjected to an Ion Exchange Strengthening Set-up," U.S. Patent 3,356,477 (December 5, 1967).

55. F.J. Shonebarger, "Strengthening Borosilicate Glass by Crowding Surface Layer With LiOH and/or KOH," U.S. Patent 3,697,242 (October 10, 1972).

5 Lithium in Glazes and Enamels

Glazes and enamels are usually treated in the ceramic literature as completely separate subjects. Each has its own method of manufacture and each is used in a different manner. For our purposes, however, glazes and enamels will simply be regarded as vitreous materials in which the effect of lithium is essentially the same. Lithium is being increasingly used in glazes and enamels because its advantages outweigh its increased cost compared to sodium and potassium.

Glazes

Raw Alkaline Glazes. According to Richardson, because all the simple compounds of soda and potash are highly water soluble and because the insoluble compounds are high in alumina and other materials, alkali glazes are usually made by fusing the soluble alkalies and silica with one or more insoluble materials.[1] The glass or frit thus produced is then ground and used alone or with other materials as a fritted glaze. Raw alkaline glazes are those in which similar materials are ground and used without preliminary treatment. Richardson recognized that it would be advantageous to use lithium because several of its simple salts, such as the carbonate, phosphate, and fluoride, are only slightly soluble in water. The goal of this study was to develop a practical, dependable glaze maturing at cone 06 that would retain the rich colors of true alkaline glazes and possess a smooth satiny surface, dull or matt. The author concluded that it is possible to make satisfactory raw alkaline glazes in which all the ingredients are insoluble or only slightly soluble and that characteristic alkaline colors can be obtained using lithium carbonate. These colors can be obtained in glazes containing from 10% to 12% lithium carbonate.

Glaze for High-voltage Porcelain. Maslennikova and Kochetkova, recognizing that the mechanical strength of glazed porcelain ware depends mainly on the thermal expansion coefficients of body and glaze, studied the use of lithium carbonate to reduce expansion.[2] They reasoned that a glaze possessing a lower thermal expansion than porcelain would promote the formation of tangential compressive stresses on the surface of the glazed ware and increase its strength. They found that the addition of the Li^+ ion, having a small ionic radius, increased the number of the more polarizable 0^{2-} ions, which was accompanied by intensification of the reaction ability of the melt. As a result of the addition of 2–4% Li_2O, a high-quality glaze coating was obtained that increased the mechanical strength of the porcelain by 13–15%. The glaze batch composition was: Lyangar feldspar 45, Tashlin quartz sand 25, Melekhov dolomite 13.5, Prosyanov kaolin 12, GO alumina 1.5, and lithium carbonate 3 weight percent.

Aventurine Glazes. Aventurine glazes are produced by adding a sufficient amount of a metallic oxide to the glaze so that it precipitates from the molten mass on cooling. Haldeman substituted lithium oxide for sodium oxide in the following composition in an effort to reduce the viscosity of the glaze and promote the formation of aventurine crystals of chromic oxide: Na_2O 0–1, Li_2O 0–1, Al_2O_3 0.1, Fe_2O_3 0–0.75, B_2O_3 1.5, Cr_2O_3 0–0.75, SiO_2 4. He concluded that, while lithium is actually detrimental to the development of aventurine crystals, it does produce a microcrystalline structure, and, in combination with chromic oxide, excellent matts.[3]

Opaque Glazes. Hummel et al. studied a combination of the systems Li_2O-TiO_2-SiO_2 and CaO-TiO_2-SiO_2 in an effort to produce opaque glaze compositions that can be fired in the usual maturing range of vitrified white-ware bodies ($\sim$ 1250°C).[4] Due to the separation of two liquids in the melt, which form finely divided spheres on cooling, the glasses produced are extremely opaque. These dispersed spherical particles that cause the opacity range in size from 0.1 to 0.5 microns.

Leadless Glazes. The American ceramic industry coats its dinnerware products with a hard, high-temperature glaze containing lead. In most cases, the lead leached from these glazes is small and can be regarded as harmless. However, some imported dinnerware is coated with lead glazes that are not as chemically resistant as their American counterparts. These can release as much as 1,000 ppm or more of lead. Cases of lead poisoning have, however, also been traced to the glazes produced by domestic manufacturers.[5] Other problems associated with the use of lead, quite apart from the possibility of release, are those of air and water pollution.

The main function of lead oxide in a dinnerware glaze is to produce a lower melting point and to provide an active solvent for the more refractory glaze materials. This results in a fluid, homogeneous coating, which cools into a smooth glaze with a high gloss. Any substitute for lead must, of course, provide similar properties.

Lithia has been shown to be an effective substitute for lead in glazes, and, therefore, it can eliminate the potential hazard of lead poisoning. For example, Cini showed that Li_2O can replace PbO and does not affect the glaze properties.[6] The optimum amount of Li_2O is one-third of the PbO content. Orlowski and Marquis have also shown by plant trials that lithia-strontia glazes are suitable as a replacement for lead glazes.[7] The compositions of two suitable lithia-strontia glazes are shown in Table 5.1.

The combination of lithium and strontium has been successful in whiteware glazes.[8] Lithium increases brilliance while strontium contributes to fusibility. The following glaze gave good results: 0.4 K_2O, 0.12 Na_2O, 0.10 Li_2O, 0.44 CaO, 0.15 MgO, 0.15 SrO, 0.35 Al_2O_3, 0.23 B_2O_3, 3 SiO_2. The lithium can be added as lithium carbonate, lithium silicate, or other compounds such as spodumene.

Behrens has recently presented a formulary of leadless glazes, many of which contain lithia.[9] As he points out, lithium has the advantage of reducing the crazing tendency of alkaline glazes, as well as increasing the hardness and chemical resistance of the glazes containing it. Some lithia-containing glazes reported by Behrens are given in Table 5.2.

Crystalline Glazes. For maximum stability, the thermal expansion of a glaze should closely match that of the body on which it is applied. This usually is no problem because, in the past, the thermal expansions of typical ceramic bodies were quite high ($> 6 \times 10^{-6}/°C$) and finding a suitable glaze to fit them

Table 5.1
Compositions of Two Lithia-Strontia Glazes

Oxide	*Glaze 1*	*Glaze 2*
Li_2O	1.03	0.40
Na_2O	2.55	1.87
K_2O	1.29	4.09
CaO	8.43	8.25
MgO	2.05	2.01
SrO	5.34	5.22
Al_2O_3	12.22	11.96
B_2O_3	5.51	5.39
SiO_2	61.59	60.29
F_2	–	0.51

Source: H.T. Orlowski and J. Marquis, "Lead Replacements in Dinnerware Glazes," *Journal of the American Ceramic Society* 28 (1945).

Table 5.2
Lithia-Containing Leadless Glazes

Type of Glaze	*Batch Composition*	
1. Low-fire alkaline glaze		
Matt glaze, cone 06	Lithium carbonate	7.1
	Frit 54 (Pemco)	40.5
	Whiting	12.9
	Kaolin	21.7
	Flint	17.8
Bright glaze, cone 06	Lithium carbonate	8.0
	Frit 25 (Pemco)	46.4
	Whiting	11.0
	Kaolin	21.0
	Flint	13.6
Clear glaze, cone 06	Lithium carbonate	9.0
	Soda ash	30.0
	Kaolin	22.0
	Flint	39.0
2. Alkaline glazes		
Cone 4	Lithium carbonate	16.5
	Nepheline syenite	30.0
	Whiting	15.6
	China clay	5.8
	Flint	32.1
Cone 9	Spodumene	48.0
	Frit 54 (Pemco)	7.4
	China clay	3.2
	Flint	41.4
3. Black glazes		
Cone 010	Lithium carbonate	23.1
	Frit 25 (Pemco)	69.4
	Zinc oxide	2.5
	Magnesium carbonate	5.0
Cone 06	Lithium carbonate	8.7
	Frit 25 (Pemco)	50.0
	Wollastonite	13.8
	China clay	20.5
	Flint	7.0
4. Bristol glazes		
Cone 04	Lithium carbonate	7.0
	Frit 3223 (Ferro)	37.2
	Zinc oxide	18.6
	China clay	18.6
	Flint	18.6

Table 5.2 (continued)

Type of Glaze	*Batch Composition*	
Cone 1 (opaque smooth matt)	Lithium carbonate	6.7
	Frit 3223 (Ferro)	34.2
	Zinc oxide	10.7
	China clay	22.0
	Flint	26.4
Cone 4 (smooth, light opaque)	Lithium carbonate	10.0
	Zinc oxide	22.0
	Whiting	4.6
	Kaolin	17.7
	Flint	38.3
	Titanium dioxide	7.4
Cone 6 (opaque mottled)	Lithium carbonate	8.7
	Zinc oxide	9.3
	Nepheline syenite	25.9
	Kaolin	11.3
	Calcined kaolin	9.7
	Flint	35.1
5. Celadon glazes		
Cone 010	Lithium carbonate	7.5
	Frit 14 (Hommel)	62.6
	Kaolin	19.7
	Flint	10.2
Cone 04	Spodumene	33.6
	Lepidolite	18.2
	Frit 14 (Hommel)	48.2
Cone 7	Spodumene	25.6
	Lepidolite	13.8
	Frit 14 (Hommel)	13.8
	Whiting	7.0
	Barium carbonate	17.2
	Flint	22.6
Cone 9	Spodumene	24.5
	Lepidolite	13.6
	Frit 14 (Hommel)	8.8
	Whiting	8.1
	Barium carbonate	16.6
	Flint	28.4
6. Glazes for cone 018–010		
Cone 018–010	Lithium carbonate	9.9
	Frit 259 (Hommel)	43.6
	Frit 14 (Hommel)	26.3
	China clay	6.2
	Flint	14.0

Table 5.2 (continued)

Type of Glaze	*Batch Composition*	
Cone 016–014	Lithium carbonate	10.4
	Frit 3223 (Ferro)	72.5
	Strontium carbonate	5.2
	Zinc oxide	2.8
	China clay	9.1
7. Majolica glaze		
Cone 014	Lithium carbonate	12
	Frit 25 (Pemco)	47
	Frit 54 (Pemco)	29
	Flint	10
	Bentonite	2
8. Stoneware glazes		
Cone 04	Lithium carbonate	10.2
	Whiting	9.8
	Barium carbonate	5.9
	Zinc oxide	9.8
	Kaolin	22.3
	Flint	42.0
Cone 1 (matt glaze)	Lithium carbonate	1.6
	Nepheline syenite	42.1
	Barium carbonate	17.0
	Whiting	10.5
	Zinc oxide	8.6
	Flint	20.2
Cone 4	Lithium carbonate	2.8
	Potash feldspar	44.1
	Wollastonite	18.5
	Barium carbonate	14.9
	Zinc oxide	2.8
	Magnesium carbonate	1.6
	Flint	15.3
Cone 6 (chocolate glaze)	Lithium carbonate	11.0
	Albany slip	85.0
	Tin oxide	4.0
Cone 9 (buckskin brown color)	Spodumene	10
	Albany slip	80
	China clay	10
9. Wide-firing range glaze		
Cone 1–6 (matt)	Lithium carbonate	5.0
	Whiting	13.5
	Frit 3293 (Ferro)	23.2
	China clay	13.9
	Calcined china clay	9.0
	Flint	35.4

Table 5.2 (continued)

Frit 54:	Na_2O 10.4; CaO 20.0; B_2O_3 23.3; SiO_2 46.3 wt. %. Molecular formula: 0.32 Na_2O; 0.68 CaO; 0.64 B_2O_3; 1.47 SiO_2.
Frit 25:	K_2O 5.4; Na_2O 14.7; CaO 0.5; ZnO 0.7; Al_2O_3 12.1; B_2O_3 16.9; SiO_2 49.7; F 1.8 wt. %. Molecular formula: 0.18 K_2O; 0.76 Na_2O; 0.03 CaO; 0.03 ZnO; 0.38 Al_2O_3; 0.78 B_2O_3; 2.65 SiO_2; 0.29 F.
Frit 3223:	Na_2O 12.3; B_2O_3 27.5; SiO_2 60.2 wt. %. Molecular formula: 1.00 Na_2O; 2.00 B_2O_3; 5.07 SiO_2.
Frit 14:	Na_2O 10.36; CaO 20.02; B_2O_3 23.26; SiO_2 46.36 wt. %. Molecular formula: 0.32 Na_2O; 0.68 CaO; 0.64 B_2O_3; 1.47 SiO_2.
Frit 259:	K_2O 6.04; Na_2O 14.14; CaO 0.27; ZnO 0.94; Al_2O_3 11.75; B_2O_3 16.50; SiO_2 48.91; F 1.45 wt. %. Molecular formula: 0.20 K_2O; 0.74 Na_2O; 0.02 CaO; 0.04 ZnO; 0.37 Al_2O_3; 0.77 B_2O_3; 2.63 SiO_2; 0.12 F.
Frit 3293:	Na_2O 16.7; CaO 0.2; MgO 0.8; Al_2O_3 5.9; SiO_2 76.4 wt. %. Molecular formula: 0.922 Na_2O; 0.012 CaO; 0.066 MgO; 0.200 Al_2O_3; 4.36 SiO_2.

Source: R. Behrens, "Glaze Projects."

was not difficult. However, recent developments of low-expansion ceramics that can be fast-fired, and thus result in lower production costs, have created a need for low-expansion glazes. Crystalline glazes based on lithia have been developed to fulfill this need.

Maki and Tashiro developed glazes that are applicable to lithia bodies having a coefficient of expansion of $< 20 \times 10^{-7}/°C$ (room temperature to $\sim 500°C$). [10] The composition of a suitable glaze for applying to a lithia-ceramic with an expansion of $5 \times 10^{-7}/°C$ is: 50.4 SiO_2, 29.2 Al_2O_3, 5.9 Li_2O, 1.7 ZrO_2, 2.6 P_2O_5, 2.6 TiO_2, 1.0 Na_2O, 1.0 K_2O, 2.8 B_2O_3, and 2.8 PbO. Suitable heat treatment converts the glaze to the crystalline state, which consists almost entirely of beta eucryptite. The chemical durability of this glaze is satisfactory for use on dinnerware.

A low-expansion, crystallizable glaze composition suitable for glass-ceramics has been patented.[11] This composition is: 45–60 SiO_2, 12–25 Al_2O_3, 16–25 Li_2O, 0–8 B_2O_3, 0–4 As_2O_3, plus a nucleating agent.

O'Conor and Eppler have recently described the preparation of low-expansion glazes compounded in the Li_2O-Al_2O_3-SiO_2 system that can be applied to bodies with thermal expansions as low as $2 \times 10^{-6}/°C$.[12] Such glazes consist of crystals dispersed in a vitreous matrix and are suitable for application to dinnerware or cookware. The compositions of the frits studied

Table 5.3
Compositions of Frits Studied

	A	*B*	*C*	*D*	*E*	*F*	*G*	*H*	*I*
Li_2O	9.1	9.0	11.3	9.0	12.8	9.0	9.0	9.0	9.4
MgO		2.0							
Al_2O_3	27.0	27.0	29.2	22.5	17.0	27.0	27.0	22.0	26.2
SiO_2	53.9	54.0	49.5	58.5	55.2	54.0	54.0	54.0	55.3
B_2O_3	5.0	4.0	5.0	5.0	7.5			5.0	2.3
K_2O	5.0	4.0	5.0	5.0	5.0	5.0	5.0	5.0	4.6
F					2.5				
CaO						5.0			
ZrO_2							5.0	5.0	2.2

Source: E.F. O'Conor and R.A. Eppler, "Semicrystalline Glazes for Low Expansion Whiteware Bodies," *American Ceramic Society Bulletin* 52 (1973).

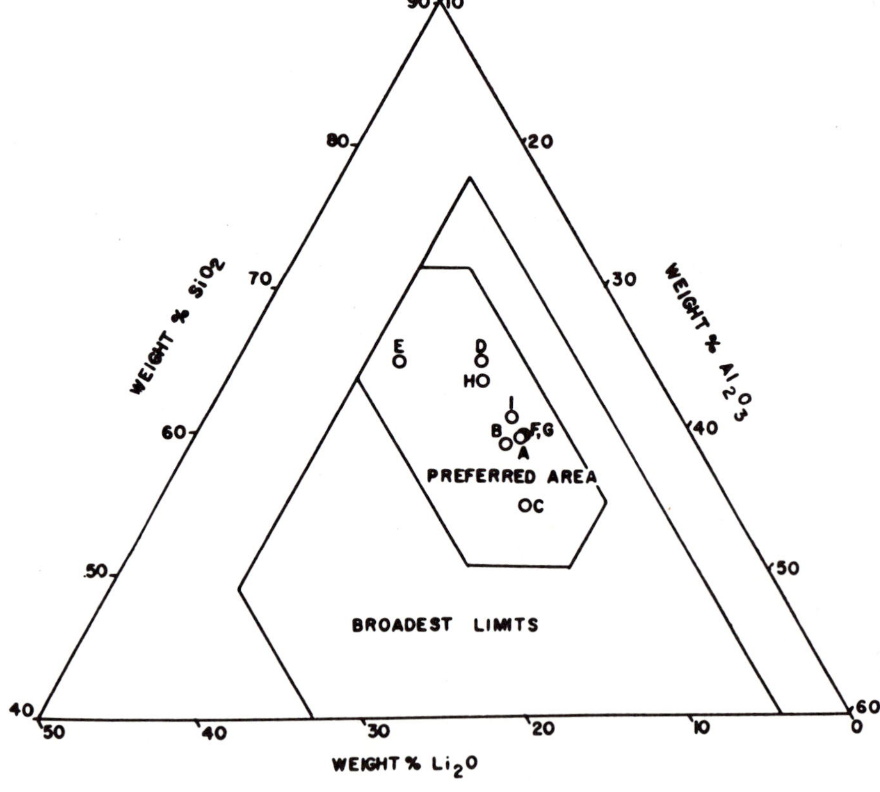

Figure 5.1. Region of the System Li_2O-Al_2O_3-SiO_2 Used in Compounding Low-expansion Glazes

are shown in Table 5.3, and the region in the Li_2O-Al_2O_3-SiO_2 system used in compounding the low-expansion glazes is shown in Figure 5.1.

Commercial Frits. Selected commercial compositions of glaze frits containing lithia are shown in Table 5.4. Pemco advertises frit P-609 as a component for partially fritted leadless glaze firing in the range of cone 4 through cone 8; the cte is 6.9×10^{-6}. P-1701 is an all-fritted glaze for high alumina bodies firing in the range 1600–1800°F; the cte is 5.4×10^{-6}. P-1855 is a very low expansion frit (cte 4.1×10^{-6}) for borosilicate glass for partically fritted glazes. P-2J57 is a phosphate-based glass and has a very high thermal expansion (cte 21.0×10^{-6}) and a low melting point (950°F–1000°F). Pb-1492 is particularly suited for wollastonite bodies maturing in the range cone 04 through cone 2; it can be used alone or in partically fritted glazes (cte 6.7×10^{-6}). The thermal expansions ($\times 10^{-6}$) and melting points (°F) of the O'Hommel frits are (in the order given in Table 5.4): 10.2 and 1230; 7.5 and 1300; 7.3 and 1750; 6.3 and 1725; and 17.6 and 1100. The Ferro frit 3292 has a cte of 7.5 and a fusion temperature of 1650°F.

Enamels

The use of lithia in enamels was first patented by D'Adrian in 1923.[13] He claimed the addition of 8–20% lepidolite to an enamel frit as a fluxing constituent of ground coat and cover coat enamels. These enamels had a minimum tendency to craze and possessed great brilliance and strength.

Lewis progressively substituted lithia, as lithium carbonate, for soda in the composition of a white cover frit and concluded that lithia resulted in improved fusibility, gloss, and opacity.[14] These enamels also exhibited higher surface abrasion and acid resistances and a lower thermal expansion than the equivalent soda enamels. From the relative fusibility curves, Lewis deduced that 1 part by weight of lithia is approximately equivalent to 2.3 parts by weight of soda.

Huppert noted that lithium often resembles the metals of Groups II and III more closely than those of Group I and suggested that the exchange of lithia for the oxides of Group II elements gives the best results.[15]

Dietzel measured the density of various glasses and found that the lithium ion (Li^+) would fit completely in the cavities of the (SiO_4^{2-}) lattice, while in the case of Na^+ and K^+, only about 70% and 27%, respectively, could be accommodated.[16] The lithium ion thus causes contraction of the glass structure and should, therefore, be suitable for making thinner enamels, which are important for improving resistance to thermal shock.

Ablard studied the effect of lithia on the strains in glass produced by

Table 5.4
Compositions of Selected Commercial Frits

Frit No.		Li_2O	K_2O	Na_2O	CaO	MgO	BaO	SrO	ZnO	Al_2O_3	B_2O_3	SiO_2	P_2O_5	F	PbO
Pemco	Wt. %	1.7	1.1	2.2	7.1	3.3	–	8.5	–	8.0	9.0	59.1	–	–	–
P-609	Mol. Wt.	0.14	0.03	0.09	0.32	0.21	–	0.21	–	0.20	0.33	2.50	–	–	–
Pemco	Wt. %	0.4	1.0	3.9	4.7	–	–	1.1	–	9.1	27.0	52.8	–	–	–
P-1701	Mol. Wt.	0.07	0.06	0.34	0.47	–	–	0.06	–	0.49	2.13	4.86	–	–	–
Pemco	Wt. %	0.3	1.5	0.3	0.1	0.1	–	–	1.0	0.9	26.2	69.6	–	–	–
P-1855	Mol. Wt.	0.31	0.30	0.09	0.02	0.04	–	–	0.24	0.16	7.19	22.3	–	–	–
Pemco	Wt. %	3.6	–	21.3	–	–	–	–	–	19.2	6.6	1.2	41.0	7.2	–
P-2J57	Mol. Wt.	0.26	–	0.74	–	–	–	–	–	0.41	0.21	0.04	0.62	0.81	–
Pemco	Wt. %	0.7	1.6	2.0	8.9	1.4	–	3.6	–	6.2	9.3	52.2	–	–	14.1
Pb-1492	Mol. Wt.	0.06	0.05	0.08	0.44	0.10	–	0.10	–	0.17	0.37	2.38	–	–	0.17
O'Hommel	Wt. %	1.01	0.74	0.25	–	0.38	–	–	–	1.45	2.75	30.87	–	–	62.50
111	Mol. Wt.	0.10	0.02	0.01	–	0.03	–	–	–	0.04	0.12	1.53	–	–	0.84
O'Hommel	Wt. %	2.26	1.96	6.47	4.06	–	2.46	–	1.37	3.81	26.13	51.49	–	–	–
361	Mol. Wt.	0.23	0.07	0.35	0.24	–	0.05	–	0.06	0.13	1.23	2.80	–	–	–
O'Hommel	Wt. %	0.41	2.96	2.79	9.73	0.72	–	4.43	–	12.12	5.26	61.58	–	–	–
389	Mol. Wt.	0.04	0.10	0.14	0.54	0.06	–	0.04	–	0.37	0.23	3.16	–	–	–
O'Hommel	Wt. %	1.02	1.29	2.55	8.45	2.07	–	5.32	–	12.21	5.48	61.61	–	–	–
390	Mol. Wt.	0.10	0.04	0.12	0.44	0.15	–	0.15	–	0.35	0.23	3.00	–	–	–
O'Hommel	Wt. %	2.70	–	21.7	1.09	–	–	–	–	18.85	6.86	5.14	40.14	3.52	–
435	Mol. Wt.	0.20	–	0.76	0.04	–	–	–	–	0.40	0.21	0.19	0.62	0.20	–
Ferro	Wt. %	0.40	3.1	3.0	10.5	0.7	–	4.8	–	10.7	5.7	61.1	–	–	–
3292	Mol. Wt.	0.042	0.095	0.141	0.535	0.055	–	0.132	–	0.301	0.235	2.92	–	–	–

applied color labels and concluded that lithia colors do not decrease the strength of bottles if their total thermal expansion is equal to or lower than the glass.[17] He also states that, because of its great fluxing action, lithia is a highly desirable ingredient of acid- and alkali-resisting ceramic colors.

According to Houston, additions of lithia have a considerable effect upon the recrystallization of titania in titania-opacified enamels.[18] Apart from producing lower firing temperatures, lithia promotes the development of rutile at the expense of anatase. Above a lithia molar concentration of 2.49%, the amount of rutile is directly proportional to the logarithm of the molar concentration of lithia and a function of the firing temperature. The overall conclusion is that, for optimum working conditions, where the ratio is anatase to rutile should be high, the amount of lithia should not exceed 1.25 mole percent.

In dry process enamels, additions of lithium carbonate decrease fritting time and increase output by 10%.[19] Lithia's strong fluxing properties produce greater uniformity, and pinholes are substantially eliminated. Lithium-containing enamels are more fluid in the molten state than those containing larger quantities of sodium or potassium.

Lithium fluoride, being a double fluxing agent because of its combination of lithium and fluorine, can be used to advantage in steel enamels, ceramic coatings, corrosion-resistant glass linings, and low-melting glasses for coating light metals.[20] When added as a mill addition to regular steel ground coats, lithium fluoride lowers and widens the firing range.

Fenton has reviewed the literature on lithium in enameling and has summarized the improvements provided by lithia as follows: [21]

1. Increase in fluidity and reduction in viscosity.
2. Reduction of maturing temperature or maturing time.
3. Increase in application rates.
4. Reduction in coating thickness.
5. Improvement in thermal shock resistance.

Acid and Alkali Resistance. Various investigators have theorized that the partial replacement of Na_2O and K_2O by Li_2O will result in improved acid and alkali resistance. This is attributed to the higher oxygen bond strength of the lithium ion in comparison with soda or potassia. For example, Staley demonstrated that lithia increased the acid resistance of wet process cast-iron enamels.[22] Again, the favorable effect of replacing soda by lithia on the alkali resistance and fusibility of enamels was shown by Vargin and Zolotova.[23] After preliminary studies, they chose an enamel of the following composition: SiO_2 65, ZrO_2 5, Na_2O 24, Li_2O 6 mole percent, and 3 parts F per 100 parts of enamel.

Simkovich and Es'kova have patented acid-resisting enamels containing lithia which exhibit increased strength and thermal stability.[24] Two compositions recommended are: SiO_2 70–78, Al_2O_3 1.5–4.5, CaF_2 1–2.5, Na_2O 4–7, Li_2O 9–12, CaO + MgO 1–5, MgO $\leqslant$ 1, P_2O_5 0.8–5.5 wt. percent and SiO_2 73.0–78.0, Al_2O_3 1.0–2.0, CaO 1.0–3.0, Na_2O 3.0–5.0, Li_2O 10.0–13.0, CaF_2 0.5–1.5, WO_3 1.0–8.0 wt. percent. Another Russian patent was issued to Khomrach and Smirnova.[25] They claimed the following range: SiO_2 58–67, Na_2O 10–14, Al_2O_3 1–3, CaF_2 1.5–6, B_2O_3 2–8, CaO 4–8, Li_2O 2–5.0, K_2O 1.5–5, CoO 0.1–0.8, Cr_2O_3 1–3, TiO_2 1–6, V_2O_5 0.2–4, and Fe_2O_3 1–4 wt. percent.

Finally, lithium is added to the compositions of glasses used for coating chemical process equipment, such as glass-lined reactors. These linings can be either glass or glass-ceramic enamels.

Thermal Expansion. The enamels Huppert developed exhibited superior resistance to thermal shock, indicating that lithium had reduced the expansion coefficient.[26] This has since been well documented. For example, Nedeljkovic and Cook studied the thermal expansion of porcelain enamels and concluded that the replacement of Na_2O by Li_2O decreased the expansion in ceria-containing compositions.[27]

Hennings points out that the expansion of all-ceramic coatings must be such that they are compatible with the body on which they are applied and that the thermal expansion of an applied glass color should not be greater than that of the glass.[28] He found that the coefficient of expansion of lithia-containing color enamels was the lowest of all compositions tested, and he listed the follow- advantages of using lithium instead of sodium in the formulation of glass color enamels:

1. Enamels can be made with improved chemical resistance and lower thermal expansion.
2. Maturing temperature can be lowered.
3. Less total alkali can be used in the batch composition.

The use of lithia-containing compounds in the production of low-expansion vitreous enamels suitable for coating borosilicate-type glasses was patented by Smith, who claimed the addition of 10–80% by weight of beta eucryptite, spodumene, or petalite.[29] The enamels of this invention are prepared by incorporating finely divided lithia-containing materials into conventional enamels or colors.

Enamels for Aluminum. Donahey et al. studied phosphate glasses as enamels for aluminum and its alloys and showed that low-temperature lithium phosphate base compositions are suitable for the formulation of lead-free enamels.[30] They demonstrated that low-fusion temperatures (below 1000°F) are obtained through the use of lithium oxide, fluorine, and boric oxide.

Bulavin reported the importance of lithia as an activator of titania opacification in silicate enamels for aluminum.[31] He showed the marked effect of Li_2O by studying the following enamel: SiO_2 28.96; TiO_2 17.0; B_2O_3 6.96; Al_2O_3 2.0; SnO_2 6.0; Li_2O 4.5; K_2O 13.08; and Na_2O 18.12. At 580°C, this enamel developed a whiteness of ∿78%. When all the lithia was replaced with soda or potassia, a clear coating was obtained even after firing up to 630°C. The general conclusion was that, to produce a leadless enamel with a high (78–80%) whiteness value, the following ratio should be used: 1 Li_2O: (1.5–2) TiO_2, and the total content of TiO_2 should be 18–21 weight percent.

Finally, Krist patented two compositions of enamel frit suitable for aluminum.[32] The frit compositions claimed are: SiO_2 34.0, Li_2CO_3 4.9, Na_2CO_3 31.0, K_2CO_3 14.5, TiO_2 20.0, Al_2O_3 2.0, V_2O_5 10.0, NaH_2PO_4 4.4, and $NaNO_3$ 2.8 parts; and SiO_2 27.3, Li_2CO_3 4.9, Na_2CO_3 32.5, K_2CO_3 13.2, TiO_2 21.0, Al_2O_3 1.5, NaH_2PO_4 6.1, Pyrobor 10.0, SrO_2 5.0, and $NaNO_3$ 3.3 parts. Both frits were melted at 1100–1150°C and quenched in water. One hundred parts of each frit were mixed with Cd red 5.0, H_3BO_3 2.0, water glass 2.0, aqueous KOH 2.0, and H_2O 45 parts, ground, and used for enameling on aluminum sheet with a 0.08 mm thick layer which was fired at 540°C.

Crystalline Enamels. Crystalline enamels suitable for coating low-expansion bodies are a recent development. Because of the necessity to produce a low-expansion crystalline phase, such enamels are usually based on the Li_2O-Al_2O_3-SiO_2 system. For example, Kosiorek and Loughman patented enamels of the following composition: SiO_2 40–67; B_2O_3 0.1–10; Al_2O_3 17–31; Li_2O 3–13; TiO_2 0.1–9; Bi_2O_3 0.1–11%.[33] These enamels are suitable for coating low-expansion glass-ceramics, and they provide beta eucryptite as the major crystalline phase.

Another patent claims the following compositional range for making crystallizable enamels for glass-ceramics: SiO_2 50–70; B_2O_3 10–25; Al_2O_3 8–18; Li_2O 1.0–10.0; TiO_2 0.1–6; Bi_2O_3 0.1–6.[34] In these enamels, beta spodumene is the primary crystalline phase produced after nucleation and crystallization.

Venzel developed glass-crystalline enamels for low carbon steel as a means for providing anticorrosion coatings for handling aggressive media.[35] He studied the crystallizability of glasses in the system diopside-spodumene-lithium disilicate. The following region of glass composition was selected: Li_2O 15–20; Na_2O 8–12; CaO 1-5; MgO 1-5; Al_2O_3 1-5; SiO_2 60–65.

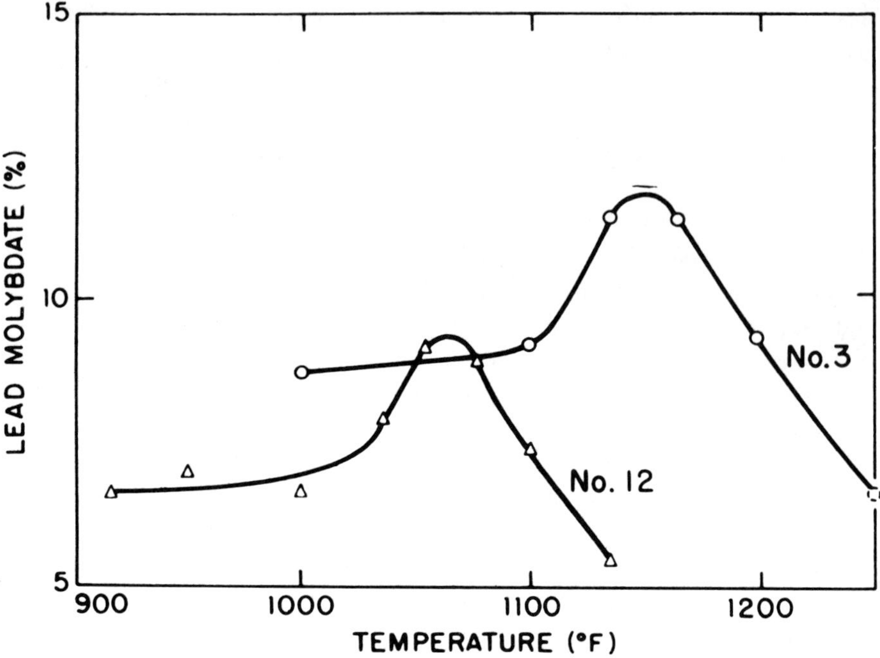

Figure 5.2. Effect of Li_2O on the Crystallization of Lead Molybdate in Compositions (3) 12.55% Na_2O, 0.0% Li_2O and (12) 10.05% Na_2O, 2.5% Li_2O. *Source:* R. B. Amin and R. L. Cook, "Crystallization Characteristics of Lead Molybdate in Low Temperature Enamels," *American Ceramic Society Bulletin* 42 (1963).

The effect of lithium on the crystallization characteristics of lead molybdate in low temperature enamels was investigated by Amin and Cook (see Figure 5.2).[36] The maximum for the amount of lead molybdate crystallized was shifted from 1150°F to 1065°F when lithia was partially substituted for soda.

Summary

Lithia is added to glassy coatings to reduce melting temperatures and increase production rates, to improve acid and alkali resistances, to increase thermal shock resistance, and to eliminate the toxic lead oxide. In crystalline glasses and enamels, the function of lithia is to produce low-expansion phases such as beta eucryptite or beta spodumene.

The particular lithia compound that should be chosen will depend on many factors, such as the desired final composition. For example, lithium carbonate is generally used in porcelain enamels because of the relatively low alumina content; this precludes the use of much spodumene. In ground coats, where iron is not a problem, the less expensive forms of spodumene concentrate can be used, whereas cover coats may require the purer grades. A low-iron spodumene is available (0.1% Fe_2O_3) for applications where the tolerance for iron is low, but which can accept the alumina values.

Notes

1. F.W. Richardson, "Use of Lithium Carbonate in Raw Alkaline Glazes," *Journal of the American Ceramic Society* 22 (1939): 50–53.

2. G.N. Maslennikova and N.F. Kochetkova, "Effect of Additions of Certain Oxides on the Properties of Glazes for High Voltage Porcelain," *Steklo i Keramika* (1970): 28–30.

3. V.K. Haldeman, "Aventurine Glazes," *Journal of the American Ceramic Society* 7 (1924): 824–833.

4. F.A. Hummel, T.Y. Tien, and K.H. Kim, "Studies in Lithium Oxide Systems: VIII, Application of Silicate Liquid Immiscibility to Development of Opaque Glazes," *Journal of the American Ceramic Society* 43 (1960): 192–197.

5. "Ceramic Dinnerware Producers Use ASTM Method to Protect Consumers," *Materials Research and Standards* (November 1972).

6. L. Cini, "Effect of Introducing Lithium Oxide into a Ceramic Glaze," *Sprechsaal* 90 (1957): 455.

7. H.T. Orlowski and J. Marquis, "Lead Replacements in Dinnerware Glazes," *Journal of the American Ceramic Society* 28 (1945): 343–357.

8. "Lithium in Clear Glazes," *Ceramic Industry* 68 (February 1948).

9. R. Behrens, "Glaze Projects," *Ceramics Monthly Magazine Handbook* (Columbus, Ohio: Professional Publications, Inc., 1971).

10. T. Maki and M. Tashiro, "Studies on Glazes for Lithia Ceramics," *Journal of the Ceramic Association of Japan* 74 (1966): 89.

11. Britain Patent 1,239,328 (1971).

12. E.F. O'Conor and R.A. Eppler, "Semicrystalline Glazes for Low Expansion Whiteware Bodies," *American Ceramic Society Bulletin* 52 (1973): 180–184.

13. A.L.D. D'Adrian, "Enamel and Method of Making the Same," U.S. Patent 1,443,813 (1923).

14. M.O. Lewis, "Effects of Lithia Substitution for Soda in Vitreous Enamel," *Journal of the American Ceramic Society* 26 (1943): 77–83.

15. P.A. Huppert, "Lithia in Porcelain Enamels," *American Ceramic Society Bulletin* 38 (1959): 57–60.

16. A. Dietzel, *Ueber die Struktur von Silikatglaesern* (Kaiser Wilhelm Institut Fuer Silkutforschung, November 1942).

17. J.E. Ablard, "Strains in Glass Produced by Applied Color Labels," *Journal of the American Ceramic Society* 28 (1945): 189–195.

18. M.D. Houston, "Influence of Lithia upon the Recrystallization of Titania," *American Ceramic Society Bulletin* 36 (1957): 139–141.

19. "Use of Lithium Carbonate in Dry Process Enamels," *Ceramic Industry* 46 (1946): 70.

20. Huppert, "Lithia in Porcelain Enamels."

21. W.M. Fenton, "Lithium as a Tool in Enameling—a Review," *Journal of the Canadian Ceramic Society* 32 (1963): 16–21.

22. H.F. Staley, "Some Relations of Composition to Solubility of Enamels in Acids," *Journal of the American Ceramic Society* 4 (1921): 703–717.

23. V.V. Vargin and I.N. Zolotova, "Alkali-Resistant Enamels," *Steklo i Keramika* 19 (1962): 23–26.

24. Z.I. Simkovich and N.F. Es'kova, "Acid-Resisting Glass-Ceramic Enamel," *U.S.S.R.* 369,107 (February 8, 1973) and their "Acid-Resisting Enamels," *U.S.S.R.* 358, 283 (November 3, 1972).

25. F.P. Khomrach and N.D. Smirnova, "Acid-Resisting Enamel," *U.S.S.R.* 357,175 (October 31, 1972).

26. Huppert, "Lithia in Porcelain Enamels."

27. A.I. Nedeljkovic and R.L. Cook, "Thermal Expansion of Porcelain Enamels Containing Cerium Oxide," *American Ceramic Society Bulletin* 50 (1971): 929–932.

28. R.A. Hennings, "Decorated Beverage Bottles," *Glass Industry* (June 1964): 289–292, 327.

29. D.C. Smith, "Vitreous Enamel," U.S. Patent 2,969,293 (1961).

30. J.W. Donahey, G.J. Morris, and B.J. Sweo, "Phosphate Base Glasses as Enamels for Aluminum and Its Alloys," *Finish* (August–September 1951).

31. Yu. I. Bulavin, "Effect of Li_2O on the Whiteness of Silicate Enamels for Aluminum," *Steklo i Keram.*, 23 (1966): 23–24.

32. O. Krist, "Enamel Frit for Aluminum," German Patent 2,119,777 (November 16, 1972), and his "Enamel Frit for Aluminum," German Patent 2,119,778 (November 16, 1972).

33. R. Kosiorek and J.I. Loughman, "Crystallizable Enamels for Glass-Ceramics," U.S. Patent 3,463,647 (1969).

34. "Crystallizable Enamels for Glass-Ceramics," U.S. Patent 3,428,466 (1969).

35. L.I. Venzel, "Glass-Crystalline Enamel Based on Lithium Metasilicate," *Steklo i Keram.* (1963): 19–22.

36. R.B. Amin and R.L. Cook, "Crystallization Characteristics of Lead Molybdate in Low-Temperature Enamels," *American Ceramic Society Bulletin* 42 (1963): 703–707.

6 Lithium in Glass-Ceramics

The development of glass-ceramics by S.D. Stookey has been said to have ushered in the fourth golden age of glass.[1] It opened the door to a whole new field of ceramic research on the study of the crystallization of glasses. Over the past several years, the ceramic literature has been filled with papers on the crystallization of glasses in a wide variety of systems. Most of these systems that yield practical glass-ceramics are based on lithia.

The fact that common or soda-lime glasses recrystallize under heat treatment is not new. Glass has long been known to be thermodynamically unstable at room temperature. It can be recrystallized below the liquidus temperature to yield "devitrite," or $Na_2O.3CaO.6SiO_2$. However, normal, uncontrolled recrystallization produces large crystals that result in an extremely weak body of no practical value. Controlled crystallization has long been used to manufacture opal and alabaster glasses, which owe their light-scattering properties to small amounts of crystalline colloidal particles.

Stookey, however, found a way to produce the ideal microstructure for a crystalline ceramic—a large number of very small, uniform, interlocking crystals packed in such a manner as to exhibit zero porosity. This microstructure can be obtained by the controlled catalyzed crystallization of glasses containing certain nucleating agents, such as titanium or zirconium oxides. These oxides are added to the initial glass batch and are melted into the glass, which is then formed into the desired shape, such as a casserole or a coffee pot. The glass articles are then heated to effect nucleation and to produce crystalline ceramics. A typical procedure for making glass-ceramics is shown in Figure 6.1. The presence of titanium or zirconium oxide is believed to promote a microliquation or glass-in-glass separation, which produces the nuclei on which subsequent crystal growth can occur.[2] The number of crystals present in the final product will be directly dependent on the number of nuclei produced, and conditions are chosen so that this number will be as large as possible.

To summarize, glass-ceramics are made by melting a suitable glass batch

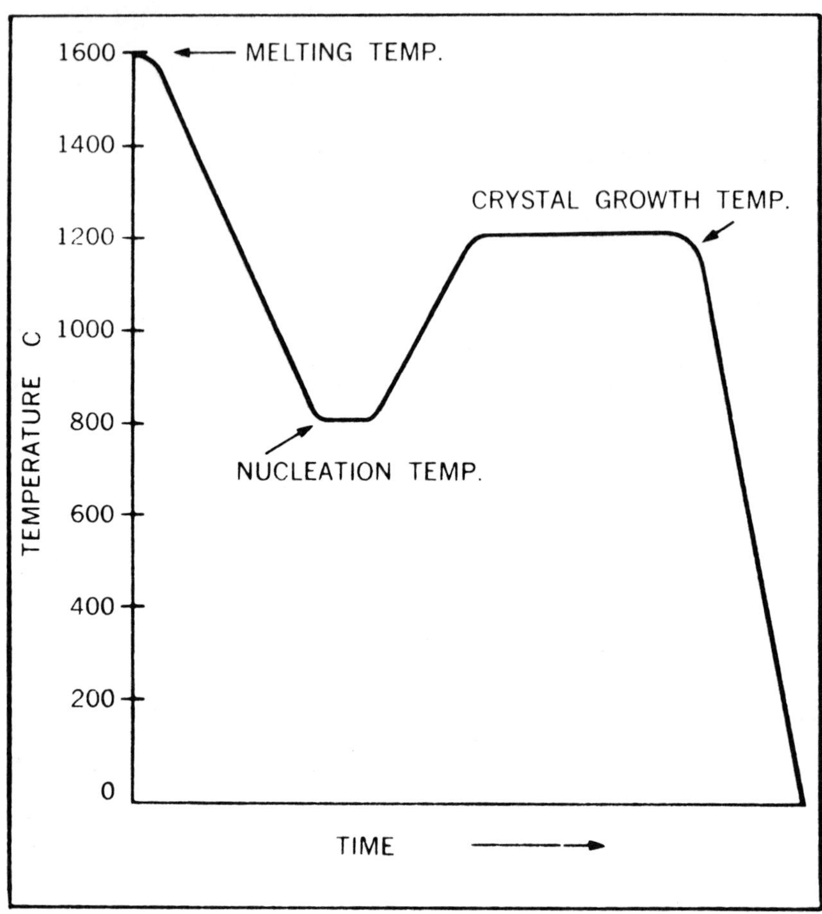

Figure 6.1. Typical Heat-treatment Curve for Making Glass-Ceramics

containing a nucleating agent, forming the glass into the desired shape, and allowing it to cool to a temperature at which nucleation occurs. This temperature is maintained until a sufficient number of nuclei has formed, after which the temperature is raised to promote crystal growth and to complete the recrystallization process.

The glass-ceramic process thus makes it possible to produce the ideal ceramic microstructure by combining the relatively low-cost process of glass manufacture with the desirable properties of a polycrystalline material. However, the practical use of these glass-ceramics is essentially dependent on the physical and chemical properties of the finished ware, and these, in turn, are a function of the nature of the crystalline phase that is produced by heat treatment.

Glass-Ceramics in the Li_2O-Al_2O_3-SiO_2 System

The first glass-ceramics to be produced commercially were sold by Corning as oven-to-freezer ware under the trade name Pyroceram®. One of the most important properties of such ovenware is its ability to withstand repeated heat shocks during transfers from oven to freezer or vice versa. This ability is based on the primary crystalline phase having an extremely low thermal expansion. This means that the ware will not be subject to great dimensional changes during heating or cooling, and no harmful stresses will be developed that could lead to failure. The crystalline phase that makes this possible is beta spodumene, or, more correctly, beta spodumene solid solution.

The low thermal expansion of lithia minerals was first reported by Hummel.[3] Later, Smoke identified two areas in the Li_2O-Al_2O_3-SiO_2 system in which the linear expansion is negative, and he developed compositions that exhibit zero expansion.[4]

Stookey was the first to recognize the importance of the Li_2O-Al_2O_3-SiO_2 system for making low-expansion glass-ceramics based on beta spodumene. This development followed logically from his earlier work on special photosensitive glass compositions containing gold, silver, or copper. Here, small proportions of light-diffusing crystals of lithium—or barium-disilicates—can be heterogeneously nucleated by exposing the glass to shortwave radiations, which produces nuclei of the photosensitve metal. Heat treatment of the glass then precipitates and grows the crystals. Compositions containing substantial amounts of lithia, resulting in the production of beta spodumene as the primary crystalline phase, were patented by Stookey in 1960.[5] The resulting ceramics possessed very low thermal expansion coefficients. Examples of eight lithia-containing compositions, taken from table II of this patent, are shown in Table 6.1.

In a later patent, Stookey concentrated on compositions containing only Li_2O, Al_2O_3, SiO_2, and TiO_2 for making crystalline ceramics with expansion coefficients below $15 \times 10^{-7}/^{\circ}C$.[6] Some compositions possessed negative expansions. Also, in certain cases, the produce was transparent even though it had a substantial crystal content. The primary crystalline phase of these glass-ceramics is beta spodumene or beta eucryptite. The range of oxide composition claimed is: SiO_2 55–75, TiO_2 3–7, $SiO_2 + TiO_2$ 58–82, Li_2O 2–15, and

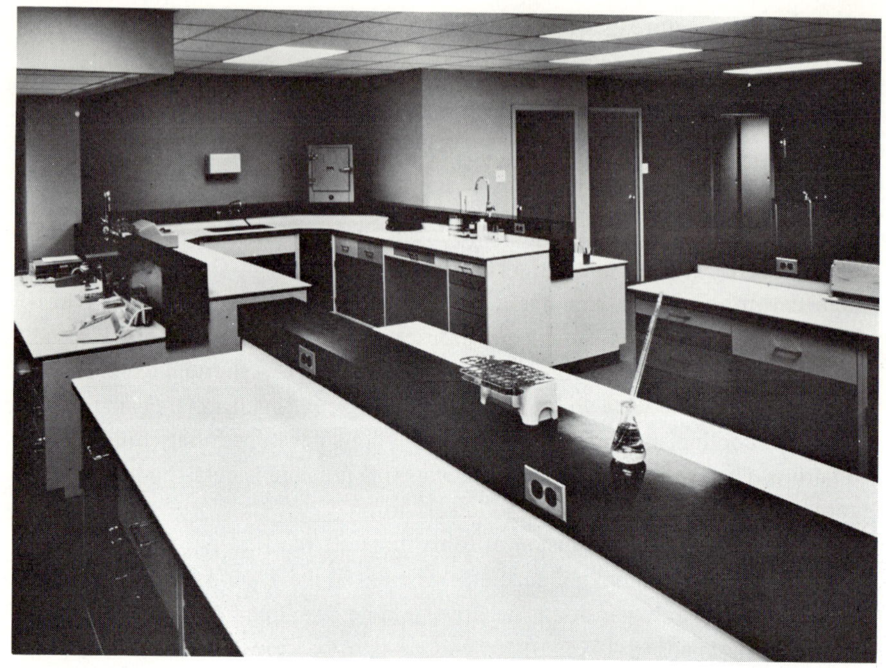

Corning labtop made from lithium aluminosilicate glass-ceramic

Corning Ware products made using low-iron spodumene

Table 6.1
Eight Lithia-Containing Glass-Ceramic Compositions

	1	*2*	*3*	*4*	*5*	*6*	*7*	*8*
SiO_2	69.8	57.6	61.7	73.1	58.7	56.1	67.4	63.7
Al_2O_3	14.9	15.2	15.3	13.5	13.7	12.1	14.4	12.1
TiO_2	7.0	12.1	10.7	4.5	13.7	13.8	7.0	13.8
Li_2O	4.3	5.2	4.3	4.9	3.9	3.0	4.2	3.1
MgO					8.8	3.9	1.2	3.9
CaO						11.1	1.8	3.4
CaF_2				2.1				
Na_2O	1.0	1.6	0.9	1.7	1.0		1.0	
K_2O			0.1	0.2	0.2			
B_2O_3	3.0	4.7	3.1				3.0	
ZrO_2		3.6	3.9					
Glass:								
Exp. Coeff./°C. × 10^7		47.9	53.5					
Density, g./cc		2.52	2.48					
Cryst'n. temp., °C	950		900		950	950	950	900
Hrs. @ cryst'n. temp.	2		2		0.5	0.5	2	1
Properties of Ceramic Products:								
Exp. Coeff./°C. × 10^7	11.4	22.0	17.2					64.4
Density, g./cc		2.57	2.55					
Strength, p.s.i. × 10^{-3}	16.3							17.8

Source: S.D. Stookey, "Method of Making Ceramics and Product Thereof," U.S. Patent 2,920,971 (January 12, 1960).

Al_2O_3 12–36 weight percent; the weight ratio of Li_2O/Al_2O_3 is 0.1 to 0.6, and the total SiO_2, TiO_2, Li_2O, and Al_2O_3 is at least 95%. The glasses are converted to ceramics by heating at 650–800°C to initiate crystallization and heating at 800°C–1175°C for one to four hours to complete the recrystallization process. Fourteen compositions based on the Li_2O-Al_2O_3-SiO_2-TiO_2 system are shown in Table 6.2.

The ultimate expansion coefficient of the ceramic depends on the heat treatment and is shown as a range in Table 6.2. The primary crystalline phases are either beta spodumene or beta eucryptite. Stookey points out that the given ranges of SiO_2, Li_2O, Al_2O_3, and TiO_2 are critical; an excess of SiO_2 or Al_2O_3 or a deficiency of Li_2O results in poor melting, a deficiency of SiO_2 or an excess of Li_2O produces a poor chemical durability, an excess of Li_2O and a deficiency of Al_2O_3 gives too high a thermal expansion, and an excess or deficiency of TiO_2 causes too rapid or incomplete recrystallization of the glass, respectively. Corning code 9608 Pyroceram, which is the basis for the oven-to-freezer ware, contains beta spodumene, cordierite, and rutile as the major crystalline phases.

Another Corning patent was issued to Voss, who claimed a composition consisting essentially of about 71 SiO_2, 2.5 Li_2O, 18 Al_2O_3, 4.5 TiO_2, 3 MgO, and 1% ZnO.[7] This glass produced a semicrystalline body with a modulus of rupture of at least 13,000 psi and a low linear thermal expansion coefficient.

Smith patented articles comprising a mixture of crystalline material and a thermally devitrifiable glass which will convert, at least partially, to the same crystalline phase as the crystalline material.[8] The crystalline material consists of 50%–95% by weight of a lithia aluminosilicate, and the glass consists of 50–55% by weight of a composition comprising Li_2O, Al_2O_3, and SiO_2. The crystalline lithia aluminosilicate used by Smith was petalite which contained 76.2–77.8 SiO_2, 16.8–17.2 Al_2O_3, 4.3–4.6 Li_2O, and 0.05–0.2% Fe_2O_3. The process patented by Smith represents a mixture of the more traditional technique of forming and firing with the new technique of controlled devitrification. Typical compositions and properties of the fired articles are shown in Table 6.3.

Reade improved upon the creep resistance of titania nucleated lithia aluminosilicate glasses by incorporating small amounts of SrO, Y_2O_3, La_2O_3, and/or Ta_2O_5 into the composition.[9] This is achieved by decreasing the viscosity of the residual glassy phase during the initial crystallization. The glass composition patented consists of 65–75 SiO_2, 15–20 Al_2O_3, 3–6 Li_2O, and 1–6 weight percent of at least one metal oxide selected from the group consisting of 1–4% SrO, 1–4% Y_2O_3, 1–5% La_2O_3, and 1–5% Ta_2O_5, and 2–7% RO_2, where RO_2 is 2–6% TiO_2 and 0–3% ZrO_2.

Table 6.2

Fourteen Glass-Ceramic Compositions Based on Li_2O, Al_2O_3, SiO_2, and TiO_2

Oxide	*1*	*2*	*3*	*4*	*5*	*6*	*7*	*8*	*9*	*10*	*11*	*12*	*13*	*14*
SiO_2	73.5	69.5	65.5	65.5	61.3	53	54.5	68	68.2	67.8	66.8	68	70.7	70.7
TiO_2	6	5.5	4.5	4.5	5	7	5.5	5	5.7	5.5	5.5	6	4.8	4.8
Li_2O	4.3	7.5	9	4	7.7	14	5.5	4.5	3.8	3.9	5.5	4	2.6	2.6
Al_2O_3	16.2	17.5	21	26	26	26	34.5	20.5	21.4	20.8	20.2	21	18.1	18.1
Expansion of glass $\times 10^7$	42	59.9	66.6	38.0	60.9	85.6	46.7	43.1	42.3	40.1	44.5	42.5	33.9	33.6
Expansion of ceramic $\times 10^7$	−4.6 to −0.7	10.3 to 12.7	14.5	−5.7 to 5.3	5.2 to 8.6	−7.7	1.1 to 12.8	−9.8 to 2.8	−3.5 to 6.9	−3.9 to 2.7	−5.3 to 7.6	−3.8 to 8.5	8.7 to 5.1	6.9

Source: S.D. Stookey, "Low-Expansion Glass-Ceramic and Method of Making It," U.S. Patent 3,157,522 (November 17, 1964).

Table 6.3
Typical Compositions Patented by Smith

	1	*2*	*3*	*4*	*5*	*6*	*7*	*8*	*9*	*10*	*11*
SiO_2	76.8	76.5	76.1	75.7	75.4	75.0	74.7	74.3	74.0	73.6	73.2
Al_2O_3	17.4	17.4	17.4	17.5	17.5	17.5	17.5	17.5	17.4	17.6	17.6
Li_2O	4.5	4.4	4.3	4.2	4.1	4.0	4.0	3.9	3.8	3.7	3.6
TiO_2		0.2	0.5	0.7	0.9	1.2	1.4	1.7	1.9	2.1	2.4
MgO	0.2	0.3	0.4	0.6	0.7	0.8	0.9	1.0	1.2	1.3	1.4
Impur.	1.1	1.2	1.3	1.3	1.4	1.5	1.5	1.6	1.6	1.7	1.8
Exp. coeff. $\times 10^7$, °C	-4.9	-4.0	-3.1	-2.2	-1.3	-0.4	+0.5	+1.4	+2.3	+3.2	+4.1
Firing temp., min (°C.)	1,310	1,290				1,220					1,170
Max. (°C.)	1,330	1,320				1,260					1,230
Opt. (°C.)	1,320	1,305				1,240					1,200

Source: G.P. Smith, "Method of Making Ceramic Article," U.S. Patent 3,246,972, April 19, 1966.

Typical applications for Corning's Pyroceram materials include heat-shock-resistant cookware for over-to-freezer use, range tops, catalyst supports, laboratory bench tops, heat exchangers, radomes, and building cladding.

Owens-Illinois has patented a transparent low-expansion crystallized glass-ceramic consisting of the following major components: 56–68 SiO_2, 18–27 Al_2O_3, and 3.4–4.5 weight percent Li_2O.[10] This glass is nucleated by 2–6 weight percent TiO_2 +ZrO_2. This composition is the basis for the Cer-Vit® glass-ceramic that has been used in a far ultraviolet scanning spectrometer aboard the Apollo 17 flight. This same lithia-based material is used to produce mirror blanks for some of the world's largest telescopes and also finds application in the manufacture of kitchen range tops.

PPG Industries manufactures a family of lithium-based glass-ceramic products called Hercuvit®. Potential uses for this material include kitchen ranges, self-cleaning oven door windows, infrared space heater covers, supersonic aircraft glazing, and laboratory bench tops. Hercuvit 106 is a translucent low-expansion product, and Hercuvit 101 is a transparent low-expansion material. Property data for both of these materials, obtained from the PPG Product Development Department, are given in Table 6.4.

Anchor Hocking Glass Corporation patented lithia-containing glass-ceramics for such use as thermal shock-resistant cookware.[11] The compositions claimed contain 2.5 to 5 weight percent Li_2O and 8–14 weight percent of a flux which includes both B_2O_3 and MgO and/or ZnO. Seven typical compositions, and the thermal expansion of the final ware, are shown in Table 6.5.

It should be noted that the thermal expansion of lithia glasses prior to crystallization is usually considerably lower than that of glasses normally used for containers. In the Anchor Hocking patent, for example, the expansion coefficient of one of the glasses is $40.2 \times 10^{-7}/°C$.

While most practical glass ceramics are based on the lithia-alumino-silica system, the basic system lithia-silica has been extensively examined for the purpose of studying the development of fine grained polycrystalline microstructures. For example, Rindone studied the crystallization of lithium silicate glasses nucleated by platinum.[12] He reported that the maximum rate of lithium disilicate crystallization was obtained using 0.005% Pt, and that an additional phase, silica O, was found when the glasses were nucleated with 0.025% Pt. Jaccodine calculated an activation energy of 49 kcal per mole for the growth process of 30 mole % Li_2O-SiO_2 glasses.[13] Freiman and Hench found that the crystallization of lithium disilicate from glasses in the Li_2O-SiO_2 system occurred through the nucleation and growth of rods.[14]

Eppler studied glass formation and recrystallization in the lithium metasilicate region of the system Li_2O-Al_2O_3-SiO_2 and concluded that the smaller the

Table 6.4
Property Data for Hercuvit 106 and 101

Hercuvit 106	
Thermal Properties	
Coefficient of thermal expansion	
$\alpha \times 10^7/°C$ (0–300°C)	-3
Specific heat, cal/g/°C	0.2
Thermal conductivity, cal/cm/sec/°C	0.004
Thermal diffusivity, cm^2/sec	0.008
Thermal shock–ΔT to ice water	
(cane sample)	927°C (1700°F)
Max. service temp.–continuous	704°C (1300°F)
short term	816°C (1500°F)
Mechanical Properties	
Density, g/cc	2.495
Modulus of rupture (abraded cane), psi	1.9×10^4
Hardness, Knoop (200 g. loading)	540
Young's modulus, psi	12.13×10^6
Rigidity modulus, psi	4.80×10^6
Bulk modulus, psi	8.46×10^6
Poisson's ratio	0.262
Optical Properties	
Index of refraction–ND	1.532

Hercuvit 101	
Thermal Properties	
Coefficient of thermal expansion	
$\alpha \times 10^7/°C$ (0–300°C)	0
Specific heat, cal/g/°C	0.20
Thermal conductivity, cal/cm/sec/°C	0.004
Thermal diffusivity, cm^2/sec	0.008
Thermal shock–ΔT to ice water	
(cane sample)	800°C (1470°F)
Max. service temp.–continuous	704°C (1300°F)
–short term	815°C (1500°F)
Mechanical Properties	
Density, g/cc	2.50
Modulus of rupture, (abraded cane) psi	1.3×10^4
Hardness Knoop (200 g. loading)	540
Young's modulus, psi	13.37×10^6
Rigidity modulus, psi	5.34×10^6
Bulk modulus, psi	8.97×10^6
Poisson's ratio	0.252

Electrical Properties	
Dielectric constant,	
25°C, 60 cps	7.6
25°C, 100 cps	7.5
25°C, 1 mc	6.9
250°C, 1 mc	9.1
350°C, 1 mc	15.8
Dissipation factor,	
25°C, 60 cps	0.034
25°C, 100 cps	0.030
25°C, 1 mc	0.006
250°C, 1 mc	0.46
350°C, 1 mc	3.73
Volume resistivity, ohm-cm,	
25°C	6×10^{12}
100°C	6×10^{11}
200°C	3×10^{7}
Surface resistivity, ohm/sq.,	
25°C	5×10^{13}
100°C	3×10^{12}
200°C	5×10^{10}
Chemical Properties	
Average surface erosion (inches)	
– 5% HCL (24 hrs.–95°C)	3.7×10^{-5}
– 5% NaOH (24 hrs.–95°C)	8.1×10^{-4}
– Water (24 hrs.–95°C)	2.0×10^{-5}

Optical Properties	
Index of refraction–ND	1.540
–1014.0 mu	1.530
Stress optical coefficient, mu/cm/kg/cm^2	3.03
Electrical Properties	
Dielectric constant,	
25°C, 100 cps	11.7
25°C, 1 mc	9.1
250°C, 1 mc	14.2
450°C, 1 mc	39.8
Dissipation factor,	
25°C, 100 cps	0.076
25°C, 1 mc	0.023
250°C, 1 mc	0.093
450°C, 1 mc	1.460
Volume resistivity, ohm-cm,	
25°C	10^{14}
250°C	10^{8}
450°C	2×10^{5}
Surface resistivity, ohm/sq., 25°C	10^{13}
(Relative Humidity 0%) 250°C	3×10^{10}
450°C	10^{10}
Chemical Properties	
Average surface erosion (inches)	
– 5% HCL (24 hrs.–95°C)	4.6×10^{-5}
– 5% NaOH (24 hrs.–95°C)	5.6×10^{-4}
– Water (24 hrs. –100°C)	4.3×10^{-5}

Table 6.5
Six Lithia-Containing Glass-Ceramic Compositions

Component	*1*	*2*	*3*	*4*	*5*	*6*	*% by weight*
SiO_2	61.24	58.98	65.1	61.24	61.26	58.78	
Al_2O_3	19.99	21.00	19.2	17.98	17.99	18.78	
Li_2O	3.42	3.29	2.87	4.42	3.42	3.28	
TiO_2	3.72	3.58	3.7	3.72	3.72	4.49	
B_2O_3	4.72	6.48	5.12	6.72	6.74	8.98	
MgO	6.31	6.08	4.6	5.31	6.32	.15	
ZnO	–	–	–	–	–	5.00	
CaO	.16	.15	–	.16	.16	.15	
Fe_2O_3	.05	.05	–	.05	.05	–	
Na_2O	.08	.08	1.06	.08	.08	.08	
K_2O	.32	.31	.24	.32	.32	.31	
Thermal expansion, per °C., $\times 10^{-7}$	19.1	19.4	16.5	13.8	17.7	12.4	

Source: "Improvements in and Relating to Semi Crystalline Glass," British Patent 1,018,384 (January 26, 1966).

percentage of lithium, the more stable the glass.[15] He found that the beta eucryptite-beta quartz solid solution is stable over most of the region. Figures 6.2 to 6.5 illustrate the occurrence of the various phases as a function of time, temperature, and constituents present, while Figure 6.6 shows the proposed liquidus relations in the system lithia-alumina-silica.

Lithium disilicate has a considerably higher thermal expansion than beta spodumene, and so it is not surprising that practical glass-ceramics based on this compound find applications quite different from those of the LAS-type. A

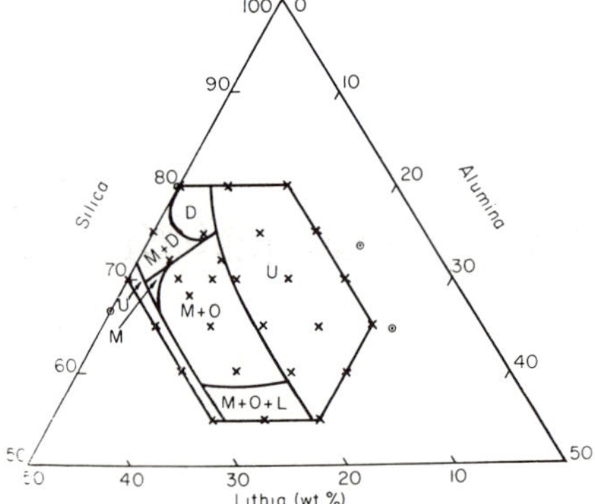

Figure 6.2. Phases Present after Crystallization of Glasses at 600°C for 69 Hours
D = lithium disilicate; L = lithium orthosilicate; M = lithium metasilicate; O = β-eucryptite-β-quartz solid solution; U = uncrystallized

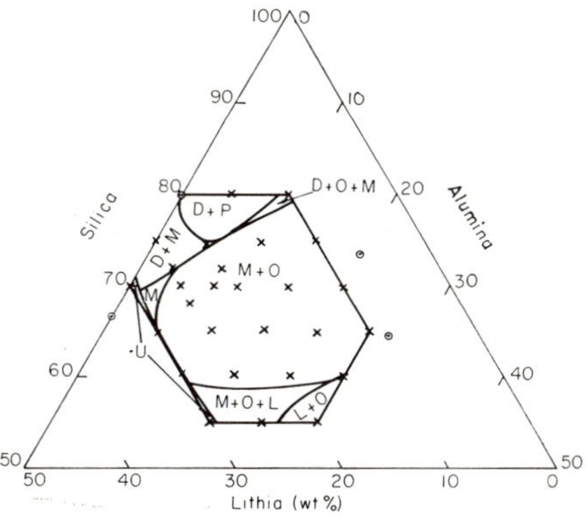

Figure 6.3. Phases Present after Crystallization of Glasses at 700°C for 69 Hours
D = lithium disilicate; L = lithium orthosilicate; M = lithium metasilicate; O = β-eucryptite-β-quartz solid solution; P = petalite; U = uncrystallized

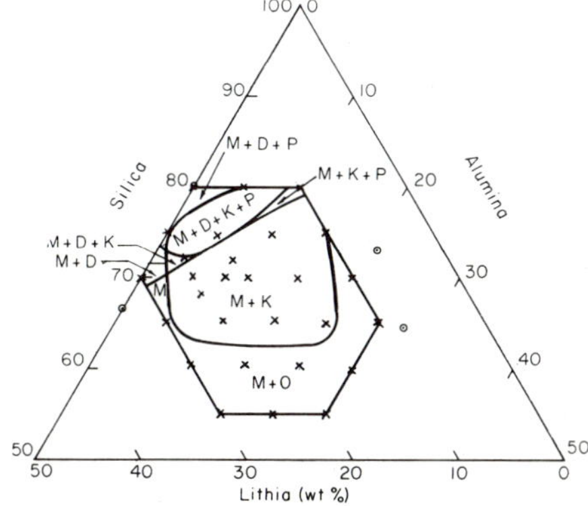

Figure 6.4. Phases Present after Crystallization of Glasses at 800°C for 69 Hours
D = lithium disilicate; K = β-spodumene solid solution;
P = petalite; M = lithium metasilicate;
O = β-eucryptite-β-quartz solid solution

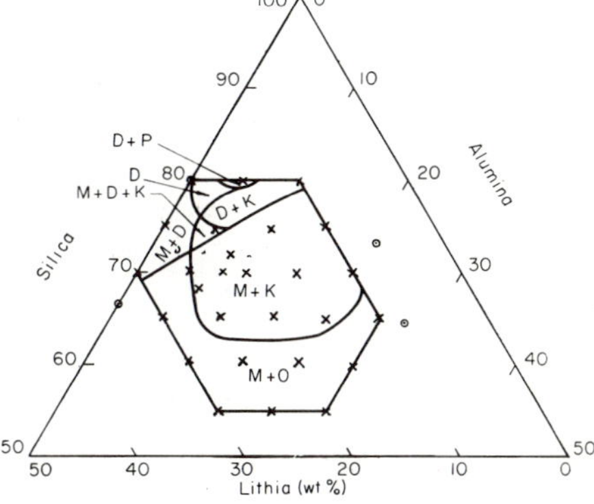

Figure 6.5. Phases Present after Crystallization of Glasses at 900°C after 69 Hours
D = lithium disilicate; K = β-spodumene solid solution;
M = lithium metasilicate; O = β-eucryptite-β-quartz solid solution;
P = petalite

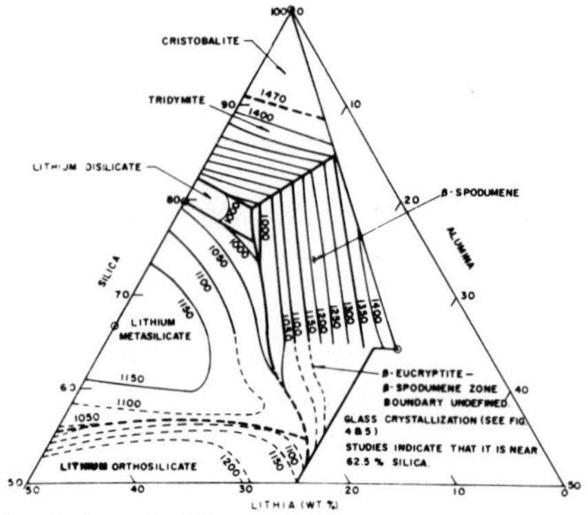

Fig. 6. The system lithia alumina–silica showing proposed liquidus relations.

Figure 6.6. The System Lithia-Alumina-Silica Showing Proposed Liquidus Relations

Table 6.6
Property Comparison of Re-X versus Conventional Porcelain

Property	*Re-X*	*Conventional Porcelain*
Modulus of rupture	25,000 psi	9–15,000 psi
Modulus of elasticity	12.6×10^6 psi	$7\text{–}14 \times 10^6$ psi
Coefficient of thermal exp. (25–800°C)	10.2×10^{-6} in/in/°C	5.0×10^{-6} in/in/°C
Dielectric strength	450 volts/mil	250–400 volts/mil
Safe temperature at continuous heat	750°C	305°C

Source: General Electric Booklet GEA 9699.

good example of the practical use of a lithium disilicate glass-ceramic is General Electric's Re-X (trademark of General Electric Company) material. Re-X is being used on capacitors, vacuum gear, transformers, and other equipment. Re-X ceramic bushings and vacuum envelopes are made by a continuous process in Pittsfield, Massachusetts. A comparison of the properties of Re-X bushings and conventional porcelain is given in Table 6.6. A detailed list of the physical and mechanical properties of Re-X glass-ceramic is given in Table 6.7, and a list of the electrical properties is given in Table 6.8.

Other Glass-Ceramic Systems Containing Lithia

According to McMillan, glass-ceramics of the Li_2O-MgO-SiO_2 system are interesting because they possess very high thermal expansions up to 140 ×

Table 6.7
Physical and Mechanical Properties of Re-X Glass Ceramic

Coefficient of thermal expansion	
0–200°C	8.6×10^{-6} in/in/°C
0–400°C	9.4×10^{-6} in/in/°C
0–600°C	10.7×10^{-6} in/in/°C
0–800°C	12.6×10^{-6} in/in/°C
Safe temperature at continuous heat	750°C
Crystal size	0.5–1 μ
Modulus of rupture	25,000 lbs/in^2
Porosity	0.00%
Density	242 gm/cc–ceramic
Hardness DPH 100 g load	630 Kg/MM
Compressive strength	150×10^3 PSI est.
Modulus of elasticity	12.6×10^6 PSI est.
Impact strength 1 cm × 1 cm × 8 cm	0.54 Ft. lbs (20°C) est.
Coefficient of friction	
Static	0.19 static est.
Dynamic	0.16 dynamic est.
Thermal conductivity	0.007 cgs units est.
Specific heat	0.22 cal/gm/°C
	0.28 cal/gm/°C
Outgassing schedule for tubes, gaps	750°C–1 hr. at 10^6 Torr
Weathering resistance	Better than glass, as good as glazed porcelain

10^{-7}.[16] Metallic phosphates have been used to catalyze glasses in the following range: SiO_2 51–88; MgO 2–27; Li_2O 9–27; and P_2O_5 0.5–6 weight percent.

In the Li_2O-ZnO-SiO_2 system, the major constituents are in the range: SiO_2 34–81; ZnO 10–59; and Li_2O 2–27. Such glasses are nucleated by P_2O_5, Ag, Au, or Cu, and produce glass-ceramics having high mechanical strengths.

Berezhnoi and Ivanova synthesized photosensitive glasses in the system Li_2O-Na_2O-K_2O-ZnO-Al_2O_3-SiO_2.[17] These glasses exhibited a low-density change during crystallization at 800°C (< 1%). They also synthesized photosensitive glass-ceramics having a low coefficient of thermal expansion of 25.9–36.4 $\times 10^{-7}$/°C.

Bogdanova et al. studied the crystallization products obtained in the system Li_2O-Al_2O_3-GeO_2-TiO_2 and found that $LiAlGeO_4$ crystallized to alpha and beta eucryptite.[18] Spodumene composition glasses crystallized to a beta eucryptite-GeO_2 solid solution. The petalite composition separated into beta eucryptite and beta spodumene phases.

Table 6.8
Electrical Properties of Re-X Ceramic

Property		*Re-X Glass Ceramic*
Volume resistivity		
23°C		6.5×10^{14} Ω cm
100°C		3.9×10^{12}
200°C		1.4×10^{10}
300°C		1.1×10^{8}
Surface resistivity		
23°C 50% R.H.		9.4×10^{14} Ω/sq.
23°C 96% R.H.		9.7×10^{8}
Electric strength (0.060″ Thick Sample) – 5 Samples		
Average		432 volts/mil
Maximum		465 volts/mil
Minimum		401 volts/mil
Dielectric constant		
(23°C)	50 Hz	6.0
	1 KHz	5.9
	1 MHz	5.8
(100°C)	50 Hz	6.8
	1 KHz	6.3
	1 MHz	6.0
(200°C)	50 Hz	12.0
	1 KHz	7.9
	1 MHz	6.3
(300°C)	50 Hz	67.1
	1 KHz	17.8
	1 MHz	7.1
Dissipation factor		
(23°C)	50 Hz	0.012
	1 KHz	0.0072
	1 MHz	0.0039
(100°C)	50 Hz	0.061
	1 KHz	0.028
	1 MHz	0.0069
(200°C)	50 Hz	0.55
	1 KHz	0.17
	1 MHz	0.019
(300°C)	50 Hz	6.34
	1 KHz	1.65
	1 Mhz	0.080

Glass-Ceramic Cements

In the fabrication of multicomponent glass-ceramic structures, a cement that will form a strong bond and yet be compatible in thermal expansion coefficient is needed. Montierth invented a glass-ceramic composition consisting of a powder of a thermally devitrifiable glass and a flux containing silica fibers.[19] The cured cement consists of 60–80% SiO_2, 18–25% Al_2O_3, 2–10% Li_2O, 0–3.3% B_2O_3, and 0.8–4.6 weight percent LiF. The principal crystalline phase of the cured cement is beta spodumene solid solution. Since beta spodumene is usually the crystalline phase present in the glass-ceramic components to be bonded, the physical and chemical characteristics of the bond match those of the components.

Chemical Strengthening of Glass-Ceramics

Glass-ceramics can be chemically strengthened in much the same manner as glasses are strengthened. For example, Karstetter and Voss show how glass-ceramics containing beta spodumene solid solution can be readily strengthened by Na^+ for Li and K^+ for Li^+ exchange.[20] The exchange of a large alkali ion for the smaller lithium ion crowds all other ions in the crystalline structure and places the surface under compression. Strengths (MOR on abraded specimens) in excess of 100,000 psi were obtained. Again, Beall et al. demonstrated that most transparent and opaque stuffed beta quartz glass-ceramics can be strengthened by ion exchange either through a $K^+ \leftrightarrows Li$ reaction in a potassium salt bath or by a $2Li \leftrightarrows Mg^{2+}$ reaction in a lithium salt bath.[21] Modulus of rupture values for abraded specimens ranged from 30,000 to 160,000 psi. Bartholomew strengthened glass-ceramics containing beta quartz and Li^+ and/or Mg^{2+} by the cation exchange reaction $Li_2SO_4 + SiO_2 \leftrightarrows Li_2SiO_3 + SO_3$ at 800°C–1175°C.[22]

Raw Materials for Lithia-Containing Glass-Ceramics

Commercial glass-ceramics are white in color and thus require low-iron raw materials in order to avoid discoloration. The combination of titanium dioxide and iron oxide is known to be particularly bad in this respect. The first commercial lithia-based glass-ceramics were made with Rhodesian petalite, which has a very high lithia-to-iron ratio. In spite of its higher lithia content, domestic spodumene was not used because it contained too high an iron content. Glass-ceramics made with domestic spodumene (0.7% Fe_2O_3) exhibited a nonuniform yellow color. The 1967 United Nations sanctions on Rhodesia, however, spurred development of a process for removing most of the iron from domestic spodumene using a high-temperature chlorination method.[23] The history of this development has been described by Fishwick.[24] The resulting product (0.1% Fe_2O_3) is suitable for use as the lithia source in most commercial glass-ceramics. Where the tolerance for iron is extremely low, lithium carbonate is the preferred lithia source.

Summary

Glasses exhibiting surface crystallization are not suitable for the production of glass-ceramics. Sodium silicate glasses crystallize from the surface, whereas lithium silicate glasses are capable of volume crystallization. Most commercial glass-ceramics are based on the lithia-alumina-silica system. Those compositions in which beta spodumene or beta eucryptite is the principal crystalline phase are used for such applications as oven-to-freezer cookware, range tops, catalyst supports, laboratory bench tops, heat exchangers, radomes, building cladding, and telescope mirror blanks. Those compositions in which the higher expansion lithium disilicate is the major crystalline phase are used for making capacitors, vacuum gear, transformers, and other equipment.

Notes

1. S.D. Stookey, U.S. Patent 2,920,971 (January 16, 1960). See his "History of the Development of Pyroceram," paper presented at the Annual Spring Meeting of the Industrial Research Institute, Colorado Springs, Colorado, May 18–21, 1958. H.H. Holscher, "Hollow and Specialty Glass: Background and Challenge," *Glass Industry* 46 (1965).
2. S.M. Ohlberg, H.R. Golob, and D.W. Strickler, "Crystal Nucleation by Glass in Glass Separation," paper presented at the Sixty-third Annual Meeting of the American Ceramic Society, Toronto, Canada, April 25, 1961.
3. F.A. Hummel, "Thermal Expansion Properties of Natural Lithia Minerals," *Foote Prints* 20 (1948): 3–11, and his "Thermal Expansion Properties of Some Synthetic Lithia Minerals," *Journal of the American Ceramic Society* 34 (1951): 235–239.
4. E.J. Smoke, "Ceramic Compositions Having Negative Linear Thermal Expansion," *Journal of the American Ceramic Society* 34 (1951): 87–90.
5. S.D. Stookey, "Method of Making Ceramics and Product Thereof," U.S. Patent 2,920,971 (January 12, 1960).

6. S.D. Stookey, "Low Expansion Glass-Ceramic and Method of Making It," U.S. Patent 3,157,522 (November 17, 1964).

7. R.O. Voss, "Method of Making a Semicrystalline Ceramic Body," U.S. Patent 2,960,802 (November 22, 1960).

8. G.P. Smith, "Method of Making Ceramic Article," U.S. Patent 3,246,972 (April 19, 1966).

9. R.F. Reade, "Glass-Ceramic Articles Containing Strontia-, Yttria-, Lanthana-, and/or Tantala-Bearing Crystal Species," U.S. Patent 3,732,116 (May 8, 1973).

10. "Glass," Britain Patent 1,124,002 (August 14, 1968).

11. "Improvements in and Relating to Semi Crystalline Glass," Britain Patent 1,018,384 (January 26, 1966).

12. G.E. Rindone, "Influence of Platinum Nucleation on Crystallization of a Lithium Silicate Glass," *Journal of the American Ceramic Society* 41 (1958): 41–42, and his "Further Studies of the Crystallization of a Lithium Silicate Glass," *Journal of the American Ceramic Society* 45 (1962): 7–12.

13. R.J. Jaccodine, "Study of Devitrification of Lithium Glass," *Journal of the American Ceramic Society* 44 (1961): 472–475.

14. S.W. Freiman and L.L. Hench, "Kinetics of Crystallization in Li_2O-SiO_2 Glasses," *Journal of the American Ceramic Society* 51 (1968): 382–387.

15. R.A. Eppler, "Glass Formation and Recrystallization in the Lithium Metasilicate Region of the System Li_2O-Al_2O_3-SiO_2," *Journal of the American Ceramic Society* 46 (1963): 97–101.

16. P.W. McMillan, *Glass-Ceramics* (London and New York: Academic Press, 1964).

17. A.I. Berezhnoi and V.I. Ivanova, "Synthesis and Study of Some Thermophysical Properties of Photoglass-Ceramics in the System R_2O-ZnO-Al_2O_3-SiO_2," *Elektronnaya Tekhnika Nauchno Tekhnicheskii Sbornik Materialov, no. 4* (1971): 48–55.

18. G.S. Bogdanova, E.M. Orlova, and Z.G. Bessmertnaya, "Phase Composition and Properties of Crystallization Products of Glasses in the Titanium Dioxide-Germanium Dioxide-Aluminum Oxide-Lithium Oxide System," *Stekloobraznye Sistemy: Novye Stekla na ikh Osnove* (1971): 158–161.

19. M.R. Montierth, "Low-Expansion Glass-Ceramic Cementing Method," U.S. Patent 3,715,196 (February 6, 1973).

20. B.R. Karstetter and R.O. Voss, "Chemical Strengthening of Glass-Ceramics in the System Li_2O-Al_2O_3-SiO_2," *Journal of the American Ceramic Society* 50 (1967): 133–137.

21. G.H. Beall, B.R. Karstetter, and H.L. Rittler, "Crystallization and Chemical Strengthening of Stuffed Beta-Quartz Glass-Ceramics," *Journal of the American Ceramic Society* 50 (1967): 181–190.

22. R.F. Bartholomew, "Strengthening of Glass-Ceramics," Japan Patent 7,201,320 (January 1972).

23. J.H. Fishwick, "Treatment of Spodumene," U.S. Patent 3,394,988 (July 30, 1968).

24. J.H. Fishwick, "Spodumene–A Review," *Foote Prints* 40 (1973): 18–28.

7
Lithium in Refractories

Many industrial ceramics must perform under conditions of sudden temperature changes. If, during these temperature changes, the stress set-up exceeds the strength of the material, failure will occur. Hasselman has pointed out that, for relatively mild thermal environments, thermal stress failure can be avoided by selecting materials with high values of strength, thermal conductivity, and thermal diffusivity and low values of thermal expansion, Young's modulus, Poisson's ratio, emissivity, and viscosity.[1] For severe thermal environments, materials should be selected with low values of thermal expansion and Young's modulus and a high value of surface fracture energy combined with high microcrack densities.

The following equation for thermal shock resistance is frequently cited when different properties are being considered:

$$\text{Index of Thermal Shock Resistance} = \frac{S_t}{\alpha E}\sqrt{\frac{k}{Cd}},$$

where S_t = tensile strength
α = linear expansion coefficient
E = modulus of elasticity
C = specific heat
d = density
k = thermal conductivity

For resistance to thermal shock, the thermal expansion coefficient is the most important property of a material. The lower the thermal expansion, the better the thermal shock resistance. Ceramic materials can be arbitrarily divided into three classes depending on their coefficient of thermal expansion:

1. Low expansion: $\alpha = < 2 \times 10^{-6}/°C$
2. Intermediate expansion: $\alpha = 2\text{–}8 \times 10^{-6}/°C$
3. High expansion: $\alpha = > 8 \times 10^{-6}/°C$

For applications involving rapid heat cycles, materials should ideally be chosen from the first group, which have expansion coefficients less than $2 \times 10^{-6}/°C$. Only fused silica and the lithium aluminosilicates exhibit thermal expansions low enough to qualify in this group (although certain cordierite compositions, and aluminum titanate, zirconyl phosphate, and tantalum, niobium, and vanadium pentoxides are also reported to have low thermal expansions), and fused silica, while having a slightly lower expansion than beta spodumene, has the disadvantage of recrystallizing above 2000°F, making it undesirable for thermal cycling.

Lithia Aluminosilicate Refractories

Hummel was the first to report the unusually low thermal expansion of beta spodumene, which is obtained by calcining natural or alpha spodumene above the temperature of the irreversible transformation.[2] He discussed the use of beta spodumene and petalite for making thermal shock resistant ceramic articles. The thermal expansion coefficient of $1.9 \times 10^{-6}/°C$ that is given in Hummel's paper is nearly double the value that is generally reported today. This could have been due either to incomplete conversion or to a relatively impure starting material. In a later paper, Hummel studied the expansion properties of synthetic lithia minerals made using lithium carbonate, alumina, and potters flint.[3] He pointed out that there is a progressive shift in the "d" spacing of the most intense x-ray line—from 3.47 to 3.44—as more silica is incorporated into the beta spodumene lattice. The thermal expansions found for the 1:1:4 ($Li_2O:Al_2O_3:SiO_2$), 1:1:6, and 1:1:8 compounds are 0.9×10^{-6}, 0.5×10^{-6}, and $0.3 \times 10^{-6}/°C$, respectively. More recently, Hummel patented a method for making low-expansion, thermal shock resistant bodies based on both calcined spodumene and petalite and also on synthetic compositions.[4] Useful compositions contained from a "few percent" Li_2O to 25% Li_2O, from 30% to 82% SiO_2, and 13% to 70% Al_2O_3. The thermal coefficient of expansion of various molecular mixtures of Li_2O, Al_2O_3, and SiO_2 is shown in Table 7.1. The coefficients vary from highly negative, when beta eucryptite is the main crystalline phase, to small and positive, when beta spodumene solid solution is the main phase. The value of $1.0 \times 10^{-6}/°C$ for beta spodumene is closer to the presently accepted value and about half of that originally reported by Hummel.[5]

Smoke recognized the practical value of using lithium aluminosilicates for making bodies that would resist failure by thermal shock.[6] He reported two

Table 7.1
Thermal Expansion of Selected Lithia Compositions

	Mol. % Li_2O	*Mol. %* Al_2O_3	*Mol. %* SiO_2	*Expansion* $\times 10^{-6}/°C$ *(Room Temperature to 1000°C)*
	1	1.17	7.45	0.5
	1	1.46	6.95	0.4
	1	0.293	0.99	- 0.7
	1	0.587	1.82	- 3.8
	1	0.685	1.65	- 4.4
Eucryptite	1	1	2	- 7.0
Spodumene	1	1	4	1.0
Petalite	1	1	8	0.2

Source: F.A. Hummel, "Thermal Shock Resistant Ceramic Body," U.S. Patent Reissue 24,765 (March 1960).

areas in which the linear thermal expansion is negative. For the area in which beta spodumene is the primary crystalline phase, the linear thermal expansions range from 0 to - 0.04%.

The first commercial application of low-expansion ceramic parts made from lithium aluminosilicates was the material Stupalith made by Stupakoff Ceramic and Manufacturing Company in Latrobe, Pennsylvania. The material was produced commercially in two principal types: with zero thermal expansion and with near-zero thermal expansion. Forming was accomplished by pressing, extruding, casting, and ramming. Use temperatures up to 2400°F were claimed, making the material suitable for pickling bath containers, kiln furniture, induction heating, high temperature jigs, and wire wound coil forms. Also, application was proposed for turbine blades, jet turbines, combustion engine parts, nozzle inserts, pouring ladles, electronic insulators, and laboratory combustion boats.

Ferro Corporation introduced a product called Zero-X, which is a synthetic lithium aluminum silicate refractory suitable for operation from 1800 to 2300°F. This material has a thermal expansion coefficient of $0.07 \times 10^{-6}/°C$ and is recommended for application in metalworking, heat treating, and electrical equipment manufacturing.

Synthesis of the lithium aluminosilicates has been performed by many investigators with a view to determining the refractoriness and thermal expansion of materials possessing various molar ratios of the oxides. For example, White and Rigby prepared a series of mixes in which the composition was varied from $Li_2O.Al_2O_3.2SiO_2$ to $Li_2O.Al_2O_3.8SiO_2$.[7] All compositions exhibited a low thermal expansion after firing to 1200°C, but their refractoriness was low. One interesting phenomenon reported was that, while negative expansions were

Various refractory shapes (foam block, oil-burner combustion chamber, permanent mold, and spark-plug holder) made from beta spodumene

obtained from certain compositions when fired in a gas furnace, similar compositions fired in an electric furnace had low positive expansions. The thermal expansions reported by White and Rigby do not agree with those reported by Hummel or Smoke.[8] The conclusion is that the ultimate expansion coefficient is determined by the mineralogical, chemical, and physical properties of the raw materials used and the method of preparation and firing.

Bulavin and Medvedovskaya also studied the synthesis of lithium aluminosilicates using lithium carbonate, Prosyanov kaolin, and Lyubevets quartz sand.[9] Formation of lithium aluminosilicates, according to these investigators, starts at 900°C with beta eucryptite being the initial reaction product. Beta spodumene formation occurs at 1150–1250°C as a result of the following reaction:

$$Li_2O.Al_2O_3.2SiO_2 + 2SiO_2 \rightarrow Li_2O.Al_2O_3.4SiO_2.$$

The beta eucryptite composition, when heated to 1350°C, is converted to beta spodumene and lithium aluminate according to the reaction:

$$(2Li_2O.Al_2O_3.2SiO_2) \rightarrow Li_2O.Al_2O_3.4SiO_2 + Li_2O.Al_2O_3.$$

Arlett et al. claimed that the partial ionic substitution of Mg^{++} (ionic radius 0.65 Å) for Li^{++} (ionic radius 0.60 Å) controls the linear thermal expansion of lithium aluminosilicate-type ceramics and also results in a ceramic that has excellent dielectric properties.[10] The ceramic material claimed consists of 61.0% frit and 39.0% clay. The frit is made by mixing 0.5-9.0% MgO, 11.5-3% Li_2O, 8% Al_2O_3, and 80% SiO_2 and melting. The frit-clay mixture is fabricated into a shape and fired at 1200-1400°C.

Shortly after Hummel reported the low thermal expansion coefficient of beta spodumene, Leitten proposed and patented the use of low-expansion grogs for stopper heads.[11] Stopper heads are used for controlling the flow of metal from steel ladles. Such grogs as lithium zirconium silicate ($Li_2O.ZrO_2.SiO_2$), beta spodumene calcined to 1100°C, and petalite calcined to 1100°C were given as examples. According to Leitten, these materials possessed expansion coefficients of 2.7, 1.9, and $1.9 \times 10^{-6}/°C$, respectively.

Later, Van Cott patented low-shrinkage, low-expansion compositions based on mixtures of beta spodumene and petalite.[12] Such bodies were fired at 1300-1350°C.

More recently, Confer and McTaggart patented a ceramic that is characterized by a zero expansion coefficient ($0 \pm 0.1 \times 10^{-6}/°C$) over the temperature range 5°C-35°C, and which has an apparent porosity of not more than 1%.[13] Suggested uses are for improved surface plates, optical tooling, optical benches, and mirror support structures. The composition of this ceramic consists of 45-55 weight percent of a glass containing 70-74% SiO_2, 22-24% Al_2O_3, 4-6% Li_2O, and 0-2% other oxides, and 45-55 weight percent of another glass material composed of 68-72% SiO_2, 17-19% Al_2O_3, 4-6% TiO_2, and 2-4% Li_2O. According to the patent, the mixture is heat treated to produce the low-expansion ceramic.

Castables. Castable refractories are mixtures of refractory grogs and a binder. They are cast as a thick suspension of solids in water. The binder provides cold setting properties by forming a bond between the grog particles. Unlike conventional refractories, which develop strength and stability only after firing, castables develop these properties during the initial cure and thereafter maintain their properties over a broad temperature range.

The first recorded study of castables based on lithium aluminosilicates was a research report for Lockheed Aircraft Corporation by Sheets et al. This study was initiated by Lockheed because of a possible need for castable refractories with a lower thermal expansion coefficient than current commercial products. The best experimental composition found consisted of a mixture of 72 weight percent of calcined petalite, 3 weight percent of ball clay, and 25 weight percent of calcium aluminate cement.

Van der Beck developed a spodumene-based castable refractory using colloidal silica as a binder.[15] The recommended composition is 44.5% of 20–50 mesh beta spodumene, 33.7% of −325 mesh beta spodumene, and 21.8% of a 30% suspension of colloidal silica. Van der Beck also reported castables bonded with an 80% spodumene slip. The properties of the silica-bonded spodumene castable are given in Table 7.2. One point to notice from this table is that the strength of the colloidal silica bonded refractory does not pass through a minimum on increasing the firing temperature. Typical hydraulic castables lose strength in the temperature range where the hydraulic bond is lost and the ceramic bond has not yet formed.

Table 7.2
Properties of a Silica-bonded Spodumene Castable

Max. service temperature °F	2100–2300*
Bulk density (lbs./cu. ft.)	
Cast (air dried)	100–120
After firing at 2200°F	101–121
Linear coefficient of expansion	
25° –450°C × 10^{-6} in/in/°C	0.5
25° –1000°C × 10^{-6} in/in/°C	0.9
Cold crushing strength–psi	
Air dried	2100
After firing at 500°F	2100
1000°F	2900
1500°F	3400
2000°F	6800
2100°F	7800
2200°F	7600
2300°F	7200
Modulus of rupture–psi	
Air dried	354
After firing at 1000°F	406
1500°F	504
2000°F	1641
2200°F	1765
2300°F	1873
Linear change–%	
Air dried	0.30
After firing at 1000°F	0.32
1500°F	0.32
2000°F	0.53
2200°F	0.57
2300°F	0.60

*Temperature dependent on time and pressure.

Fishwick and Talley described the preparation of low-expansion castable refractories based on beta spodumene and calcium aluminate cement.[16] The physical properties of dense spodumene-CA 25 (a calcium aluminate cement from Alcoa) bodies with and without grog are shown in Table 7.3 and of lightweight bodies in Table 7.4.

One composition, containing 16.5% CA–25, was selected for further study. The effect of firing temperature on the physical properties of this body are shown in Table 7.5.

To illustrate the excellent thermal shock resistance of these spodumene-based castables, the modulus of rupture of bars (6 × 1 × 3/4″) was measured after thermal cycling from 2000°F. As indicated by the data in Table 7.6, no loss in strength was found after twenty-five cycles.

Applications recommended for spodumene castables include car top refractories and kiln furniture.

Petalite has been used for many years by Robinson Clay Product Company for the construction of tunnel kiln cars.[17] This product has a thermal expansion coefficient of $1.5 \times 10^{-6}/°C$ and a use temperature of 1260°C.

Gugel also studied the use of petalite compositions for car top applications, and he recommended using a lithium fire clay, made by reacting petalite with a refractory clay and adding calcined petalite.[18] Refractories with a thermal expansion coefficient of $1\text{–}1.5 \times 10^{-6}/°C$ were achieved.

Recent developments in the aircraft and other industries have created a need for dies used in the forming of metals such as titanium. Such dies must withstand relatively high temperatures without cracking or spalling. Recognizing this need, Spangler patented a ceramic made by firing a mixture of alpha spodumene, petalite, and clay in the proportion of at least 10% of spodumene, at least 10% of petalite, and 10% to 60% of clay.[19] Such a ceramic has an expansion coefficient of from -0.3 to $-0.1 \times 10^{-6}/°C$. After crushing to specified sizes, this composition can be bonded with 15% to 30% of calcium aluminate.

Foamed Refractories. With similar applications in mind, Fishwick studied the manufacture of lightweight ceramics based on petalite and beta spodumene.[20] Such products are suitable for use as porous refractories for the simultaneous brazing and heat treatment of high-temperature alloys for use in supersonic aircraft, where conventional alloys cannot be used because of the high temperatures involved. The foams were prepared by aerating a slip containing either petalite or beta spodumene in the presence of a foam stabilizer and a binder. After casting, the bodies were dried and fired. The results for this study are shown in Table 7.7.

Table 7.3
Physical Properties of Dense Spodumene–CA 25 Bodies with and without Additions of Grog

No.	% 20 mesh β-spod.	% −325 mesh β-spod.	% CA–25	% grog added	% Shrinkage at 250°F/2000°F/Total (5 hrs.)			MOR, psi Air Dried/250°F/2000°F			Exp. Coeff.* × 10^{-6}/°C	Density #/cu. ft.
					250°F	2000°F	Total	24 hrs.	5 hrs.	5 hrs.		
1	53	40	7	–	0.20	0.45	0.65	30	95	195	1.09	89
2	51	39	10	–	0.20	0.56	0.76	83	193	277	1.04	92
3	50	37	13	–	0.19	0.80	0.99	114	296	493	1.22	93
4	47.5	36	16.5	–	0.23	0.79	1.02	172	442	648	1.29	96
5	62	31	7	–	0.21	0.47	0.68	70	70	143	0.91	89
6	60	30	10	–	0.21	0.52	0.73	124	157	256	1.13	89
7	58	29	13	–	0.30	0.68	0.98	120	267	437	1.17	89
8	56	27.5	16.5	–	0.32	0.74	1.06	168	362	661	1.30	92
9	47.5	36	16.5	15	0.28	0.57	0.85	–	341	415	1.53	94
10	47.5	36	16.5	30	0.27	0.37	0.64	–	334	272	1.69	95
11	46	34	20.0	15	0.28	0.66	0.94	–	540	467	1.61	98
12	46	34	20.0	30	0.30	0.43	0.73	–	581	410	1.63	102
13	46	34	20.0	40	0.33	0.30	0.63	–	547	334	2.02	104

*Room temperature to 600°C.

Table 7.4
Physical Properties of Lightweight Spodumene-CA 25 Bodies

No.	% −325 mesh β-spod.	% CA-25	% Ball Clay	% Verilite	% 40M Sawdust Added**	% Vermiculite	% Micro-balloons added**	% Perlite	% Shrinkage at 250°F 5 hrs.	% Shrinkage at 2100°F 5 hrs.	% Shrinkage Total	MOR, psi 250°F 5 hrs.	MOR, psi 2100°F 5 hrs.	Exp. Coeff.* $\times 10^{-6}/°C$	Density lbs/cu. ft.
14	80	10	–	10	–	–	–	–	0.33	1.17	1.50	231	606	1.43	84
15	65	10	–	25	–	–	–	–	0.25	1.44	1.69	254	539	1.88	82
16	75.6	10	5	9.4	–	–	–	–	0.43	2.30	2.73	110	520	1.60	84
17	61.4	10	5	23.6	–	–	–	–	0.31	2.34	2.65	290	896	2.00	81
18	80	5	5	10	–	–	–	–	0.58	2.28	2.86	79	837	1.43	87
19	65	5	5	25	–	–	–	–	0.46	2.70	3.16	144	1026	1.95	82
20	90	10	–	–	9.6	–	–	–	0.65	0.79	1.44	89	300	1.10	66
21	85	10	5	–	9.6	–	–	–	0.93	1.14	2.07	95	355	1.23	67
22	90	10	–	–	–	–	4.2	–	0.54	1.12	1.66	100	449	1.34	66
23	85	10	5	–	–	–	4.2	–	0.72	1.62	2.34	136	593	1.44	69
24	79.7	9.4	4.6	–	–	6.3	–	–	0.70	3.23	3.93	93	1411	1.53	83
25	78.9	9.3	4.6	–	–	–	–	7.2	0.37	2.04	2.41	195	1129	1.77	76

*Room temperature to 600°C.
**Above 100%.

Table 7.5
Effect of Firing Temperature on the Physical Properties of a Spodumene-16.5% CA-25 Body

Firing or Curing Temp. °F	*Bulk Density Lbs/cu. ft.*	*Cold Crushing Strength, psi*	*MOR, psi*	*Total % Shrinkage*	*Expansion Coefficient $\times 10^{-6}/°C$**
250	96	1558	454	0.38	–
1000	95	1359	383	0.74	–
1500	94	1314	349	0.87	–
1800	95	1472	380	0.91	1.36
1900	94	2030	522	1.03	1.36
2000	95	2850	732	1.20	1.29
2100	94	3512	814	1.22	1.22
2200	97	7595	1638	1.92	1.08

*Room temperature to 600°C.

Table 7.6
Effect of Thermal Cycling on the Transverse Strength of Spodumene-16.5% CA-25 Bodies

Cycles	*MOR, psi (After 2000°F, 5 hrs.)*
0	732
5	727
10	671
25	758

Spodumene-Clay, Petalite-Clay Systems. For many forming operations, clay is an essential component. In addition to facilitating forming, clay also can increase the refractoriness of lithium aluminosilicates. Fishwick et al. suggested that the reaction between spodumene and clay should proceed as follows: [21]

$$\underset{\text{Spodumene}}{Li_2O.Al_2O_3.4SiO_2} + 3\,\underset{\text{Metakaolin}}{(Al_2O_3.2SiO_2)} \rightarrow \underset{\text{Beta Spodumene Solid Solution}}{Li_2O.Al_2O_3.8SiO_2} + 3\,\underset{\text{Mullite}}{Al_2O_3.2SiO_2}$$

Note that, while spodumene can assimilate at least four moles of silica into solid solution, petalite already contains close to the maximum tolerable amount. Calculation shows that, at about 30% spodumene, all the silica that is

Table 7.7
Properties of Foamed Ceramics Based on Petalite and Beta Spodumene

Sample No.	Composition	Cone	Color of Brick	Density g/cm³	Linear thermal exp. coeff. × 10^6 (in./in./°C) Room Temp.– 300°	Room Temp.– 600°	Crushing Strength (psi)	Modulus of Rupture (psi)	X-ray Results*
1	88%S, 10%K, 2%T	3	Pale pink	0.88	0.40	0.60	232	103	Some evidence of quartz
2	88%S, 10%K, 2%T	7	Light buff	0.99	0.48	0.92	1445	501	Less quartz than sample 1
3	88%S, 10%K, 2%T	10	Brown	1.03	0.44	0.79	4870	1052	No quartz peaks
4	78%S, 20%K, 2%T	3	Pale pink	0.86	0.52	0.96	546	278	Marked evidence of quartz
5	78%S, 20%K, 2%T	7	Light buff	1.00	0.58	1.03	4070	1021	No quartz peaks
6	78%S, 20%K, 2%T	10	Light buff	1.00	0.70	1.11	4185	1234	No quartz peaks
7	88%P, 10%K, 2%T	3	White	0.89	0.29	0.67	417	139	Very strong quartz peaks
8	88%P, 10%K, 2%T	7	White	0.88	0.66	1.41	852	292	Very strong quartz peaks <sample 7
9	88%P, 10%K, 2%T	10	White	0.93	0.36	0.65	3450	1082	Very strong quartz peaks >sample 4
10	78%P, 20%K, 2%T	3	White	0.90	0.10	0.38	556	212	Very strong quartz peaks
11	78%P, 20%K, 2%T	7	White	0.84	0.62	1.16	806	391	Very strong quartz peaks <samples 10 & 8
12	78%P, 20%K, 2%T	10	White	1.02	0.50	0.86	4435	913	Small amount of quartz

* The main phase in all samples is β-spodumene; apart from free quartz, no other compound was detected.

S = spodumene

P = petalite

K = kaolin

T = talc

exsolved as metakaolin is converted to mullite can be assimilated in the spodumene structure, thereby changing its molar ratio from 1:1:4 to 1:1:8. Various spodumene-clay and petalite-clay compositions were fired from cone 12 to cone 16, and the physical properties of the bodies were recorded. The results of this study are given in Table 7.8. X-ray diffraction patterns were obtained from several compositions, and these results are given in Table 7.9.

X-ray diffraction analysis of fired kaolin revealed mullite as the primary phase and cristobalite as the secondary. The cristobalite peaks are enhanced in samples 2 and 10, which correspond to 10% additions of spodumene and petalite, respectively. In the spodumene-kaolin system, no cristobalite was found in compositions containing more than 10% spodumene. In the petalite-kaolin system, cristobalite persisted at least up to 30% petalite. In the spodumene-kaolin system, the first appearance of beta spodumene solid solution occurred in the sample containing 30% spodumene (sample 4) in bodies fired to cone 13. An increase in the spodumene content of bodies fired to cone 13 produced a reduction in the mullite peaks and a gradual increase in the "d" values of the beta spodumene phase. The presence of beta spodumene solid solution in the spodumene-kaolin compositions was first noted at 30% spodumene for cone 13, 40% for cone 14, and 50% for cones 15 and 16. A considerable amount of the spodumene must be in a glassy phase.

In the petalite-kaolin system, no beta spodumene solid solution was found above cone 13. Once again, the formation of a glass would be responsible for this.

Very low thermal expansion coefficients can be obtained in both the spodumene-clay and petalite-clay systems when bodies made from them are fired to cones 14, 15, and 16. However, bodies that exhibit the lowest expansions are also the weakest.

The combination of very low thermal expansion and very high refractoriness has not yet been attained for ceramic materials. An attempt was made to find a compromise by making a prereacted raw material from mixtures of spodumene and kaolin and using this to fabricate bodies. The results obtained were influenced to a large extent by the method of testing. Thus, using the pyrometric cone test, low-expansion bodies ($<2 \times 10^{-6}/°C$) were obtained with PCE values in excess of cone 18. However, when the refractoriness was evaluated using hot load testing, the results were not as encouraging. The usefulness of bodies made from calcined mixtures of spodumene and clay will depend largely on the extent to which they are loaded at high temperatures.

Table 7.8
Physical Properties of Compositions in the Spodumene-Kaolin and Petalite-Kaolin Systems

	Composition			Cone 12			Cone 13			Cone 14			Cone 15			Cone 16		
Sample No.	Spodumene (%)	Petalite (%)	Kaolin (%)	Firing Shrinkage (%)	MOR (psi)	Exp. Coeff. × 10^{-6} in./in. °C	Firing Shrinkage (%)	MOR (psi)	Exp. Coeff. × 10^{-6} in./in. °C	Firing Shrinkage (%)	MOR (psi)	Exp. Coeff. × 10^{-6} in./in. °C	Firing Shrinkage (%)	MOR (psi)	Exp. Coeff. × 10^{-6} in./in. °C	Firing Shrinkage (%)	MOR (psi)	Exp. Coeff. × 10^{-6} in./in. °C
1			100	10.4	2,830	3.52	11.3	2,958	3.76	11.9	2,455	7.39						
2	10		90	8.5	6,153	8.28	8.9	6,700	7.81	11.1	7,002	7.39	12.1	8,213				
3	20		80	10.5	9,142	4.30	10.5	7,217	4.43	11.7	6,057		11.6	8,743				
4	30		70	7.9	8,264	2.83	9.0	6,718	2.39	9.3	2,107	3.15	9.2	6,930	4.2			
5	40		60	4.2	6,762	2.70	7.1	6,538	1.63	7.3	3,196	0.70	7.0	<500	2.30	6.5	<500	3.65
6	50		50	1.1	5,406	2.40	6.2	9,563	1.93	5.7	2,430	0.17	5.1	<500	0.09	5.7	<500	0.65
7	60		40		7,008		5.5	12,377	2.19	5.5	2,277	-0.19	5.7	<500	-0.42		<500	-0.24
8	70		30	-3.3*	3,685	1.97	1.7	6,525	1.88					<500				
9	90		10	0.9	4,936	1.41	5.6	10,440	1.44									
10		10	90	8.7	4,479	8.43	9.4	6,600	8.64	10.6	6,940	8.23	12.0	8,184				
11		20	80	9.5	5,221		11.0	11,416	7.87	13.1	10,113		13.6	10,543				
12		30	70	12.2	9,542	5.98	12.6	15,235	5.65	12.8	12,610	5.50	12.1	9,601	4.30		8,250	3.84
13		40	60	11.2	10,029	4.73	10.8	11,682	4.78	9.9	12,130		8.6	7,220				
14		50	50	10.0	7,941	3.90	9.1	9,259	4.19	7.9	10,280	4.10	6.2	6,203	3.91		7,120	3.74
15		60	40		10,245	0.79					6,199	2.77		3,782	2.40		1,758	2.99
16		70	30		9,575	0.61	10.1		-0.41		6,678	0.25		4,998	0.07		1,358	-0.14
17		90	10			0.50	10.3	8,573	0.59									

*Expanded

Note: All expansion coefficients are for the range room temperature to 600°C.

Table 7.9
X-ray Diffraction Data for Several Compositions in the Systems Spodumene-Kaolin and Petalite-Kaolin

Sample Number	*Cone 13*	*Cone 14*	*Cone 15*	*Cone 16*
1	M; C(m)	M; C(s)		
2	M; C(s)	M; C(s)		
3		M		
4	M; BSS	M; BQ(w)		
5		M; BSS	M; BQ(w)	M
6	M; BSS	M; BSS	M; BSS	M; BSS
7		M; BSS	M; BSS	M; BSS
8	M; BSS			
9	M; BS			
10	M; C(s)	M; C(s)		
11				
12	M; C(m)	M; C(m); BQ(w)		
13				
14	M; BSS	M; BQ(m)		M
15		M; BQ(m)	M; BQ(w)	M; BQ(w)
16	M; BSS; Q(vw)	M; BQ (m)	M; BQ(m)	M; BQ(m)
17	M; BSS; Q(w)			

M = Mullite
C = Cristobalite
BSS = Beta spodumene solid solution
BS = Beta spodumene
BQ = Beta quartz

Q = Quartz
s = strong
m = moderate
w = weak
vw = very weak

Raw Materials

For applications in which the large volume increase at transformation poses no problem, alpha or natural spodumene is the preferred raw material. While several grades of alpha spodumene are available commercially, it should be remembered that the grades containing the maximum amount of spodumene (highest lithia content) will exhibit the greatest refractoriness. The presence of feldspar and mica will lower the refractoriness of spodumene-based bodies considerably. These spodumene concentrates are typically available in two particle sizes, 20 × 140 mesh and −200 mesh.

Most applications, however, will require beta spodumene as the major raw material component. Beta spodumene has already been calcined above the transformation temperature, thus eliminating further volume changes that could lead to cracking on firing. It is available from Foote Mineral Company under the trade name Thermal Grain® in sizing of 20 mesh by down and −325 mesh.

Most of the petalite that was used for making low-expansion refractories was imported from Rhodesia. The 1967 United Nations sanctions currently prevent this material from entry into the United States.

Summary

Thermal expansion is one of the most important properties that determine the ability of ceramic ware to withstand thermal shock. Beta spodumene is one of the very few materials that exhibit a coefficient of less than $1 \times 10^{-6}/°C$, and is thus a prime candidate for the manufacture of heat-shock resistant refractories. Both dense and lightweight refractories based on beta spodumene or beta spodumene-clay have been made and patented for such applications as car top refractories, kiln furniture, and metal brazing fixtures.

Notes

1. D.P.H. Hasselman, "Thermal Stress Crack Stability and Propagation in Severe Thermal Environments," *Ceramics in Severe Environments,* ed. by W.W. Kriegel and H. Palmour III (New York: Plenum Press, 1971): 89–103.
2. F.A. Hummel, "Thermal Expansion Properties of Natural Lithia Minerals," *Foote Prints* 20 (1948): 3–11.
3. Ibid., "Thermal Expansion Properties of Some Synthetic Lithia Minerals," *Journal of the American Ceramic Society* 34 (1951): 235–239.
4. Ibid., "Thermal Shock Resistant Ceramic Body," U.S. Patent Reissue 24,795 (March 1960).
5. Hummel, "Thermal Expansion Properties of Natural Lithia Minerals."
6. E.J. Smoke, "Ceramic Compositions Having Negative Linear Thermal Expansion," *Journal of the American Ceramic Society* 34 (1951): 87–90, and his "Ceramic Compositions Having Negative Linear Thermal Expansion: Part II," *Ceramic Age* (July 1953).
7. R.P. White and G.R. Rigby, "The Thermal Expansion Properties of Compositions Containing Lithia, Alumina, and Silica," *Transactions of the British Ceramic Society* 53 (1954): 324–34.
8. Hummel, "Thermal Expansion Properties of Some Synthetic Lithia Materials"; Smoke, "Ceramic Compositions."
9. I.A. Bulavin and E.I. Medvedovskaya, "A Study of the Synthesis of Lithium Aluminosilicates," *Neorganicheskie Materialy* 5 (1969): 1435–1438.
10. R.H. Arlett, S. DiVita, and E.J. Smoke, "Methods for Controlling the Thermal Expansion Properties of Ceramics," U.S. Patent 3,309,208 (March 14, 1967).
11. C.F. Leitten, "Stopper Heads," U.S. Patent 2,772,176 (November 27, 1956).

12. H.C. Van Cott, "Method of Making a Sintered Ceramic Article," U.S. Patent 3,096,159 (July 2, 1963).

13. J.O. Confer and D. McTaggart, "Ceramic Article and Method of Making It," U.S. Patent 3,715,220 (February 6, 1973).

14. H.D. Sheets, C. Hyde, and W.H. Duckworth, "Development of a Refractory Castable Based on Calcined Petalite," *Summary Report,* Battelle Memorial Institute (December 11, 1959).

15. R.R. Van der Beck, "A Spodumene-based Castable Refractory," *American Ceramic Society Bulletin* 42 (1963): 448–449.

16. J.H. Fishwick and R.W. Talley, "Castable Refractories Based on Spodumene," *Ceramic Age* 2 (February 1966).

17. "Top Block Breakage Cut with Petalite Addition," *Brick and Clay Record* 135 (1960): 63.

18. E. Gugel, "Fire-Resistant Concrete with Low Thermal Expansion," *Sprechsaal für Keramik, Glas, Email, Silikate* 100 (1967): 825–830.

19. E.G. Spangler, "Ceramics Produced from Spodumene, Petalite, and Clay," U.S. Patent 3,690,904 (September 12, 1972).

20. J.H. Fishwick, "Manufacture of Foamed Ceramics Based on Petalite and Beta Spodumene," *American Ceramic Society Bulletin* 42 (1963): 110–113, and his "Lightweight Ceramic Product and Method of Making," U.S. Patent 3,413,132 (1968).

21. J.H. Fishwick, R.R. Van der Beck, and R.W. Talley, "Low Thermal Expansion Compositions in the Systems Spodumene-Kaolin and Petalite-Kaolin," *American Ceramic Society Bulletin* 43 (1964): 832–835.

8
Lithium in Whitewares

The principal raw materials used in making whitewares are clay, feldspar, and quartz. The clay imparts plasticity and green strength, the feldspar behaves as a flux, and the quartz is a filler. After firing, triaxial whiteware bodies typically contain mullite, quartz, and glass as the primary phases. Cristobalite may also be present in high quartz bodies. Thus, whiteware bodies consist of crystalline phases dispersed in a glassy matrix, and some pores are usually present.

Lithia can be added to whiteware bodies in varying amounts—from very small to relatively large—depending on whether it is to be regarded as a flux or as a low-expansion filler. When used as a flux, the lithia is completely dissolved in the glass phase, where it reduces the expansion and viscosity of the glass, thereby lowering firing temperatures. When used in larger amounts, some of the lithia may still be assimilated into the glass phase, but most of it is consumed in forming a low-expansion lithium aluminosilicate. This low-expansion phase markedly reduces the expansion coefficient of the whiteware body, enabling it to be very fast fired and also making it suitable for heat-shock-resistant applications.

Lithia as a Flux

Boyd studied the pyrometric properties of spodumene-feldspar mixtures and demonstrated that they have fusion temperatures as much as six to seven cones below feldspar.[1] The area of lowest PCE values has the following composition limits:

Mineral	*% Maximum*	*% Minimum*
Spodumene	35	15
Potash feldspar	50	25
Soda feldspar	50	20

Boyd used a relatively pure spodumene (7.26% Li_2O) in his experiments. Present commercial grades of spodumene concentrates usually contain somewhat less lithia than this because of the presence of small amounts of other minerals, such as feldspar, quartz, and mica. Thus, when calculating the relative amounts of spodumene and feldspar to be used for maximum fluxing ability, reference should be made to the limits expressed in chemical terms as follows:

Oxide	*Maximum (%)*	*Minimum (%)*
Li_2O	2.5	1.1
K_2O	6.2	3.2
Na_2O	5.1	3.2

Schurecht et al. examined the effect of fluxes composed of spodumene and feldspar on semivitreous bodies and in vitreous hotel chinaware bodies.[2] They found that the most fusible mixture was composed of 30% spodumene, 10% potash feldspar, and 60% soda feldspar. This mixture has a softening point of 1053°C. The recommended mixtures of spodumene and feldspar for various bodies are given in Table 8.1.

According to Schurecht et al., the main advantage gained by the use of spodumene-feldspar mixtures to replace feldspar in whiteware bodies is a lowering of the vitrification temperature and an improvement in the strength of bodies fired at lower temperatures.

An investigation by Cowan et al. showed that slips of spodumene-containing bodies are as stable as normal feldspar bodies and require no more electrolyte for cast ware.[3] They demonstrated that spodumene can be used to advantage

Table 8.1
Recommended Spodumene-Feldspar Mixtures

Body	*Flux*
Semivitreous body, cone 6	40% spodumene; 40% K_2O feldspar; 20% Na_2O feldspar
Semivitreous body, cone 8	30% spodumene; 60% K_2O feldspar; 10% Na_2O feldspar
Semivitreous body, cone 12	40% spodumene; 60% K_2O feldspar
Vitreous body, cone 6	60% spodumene; 40% K_2O feldspar
Vitreous body, cone 8	40% spodumene; 40% K_2O feldspar; 20% Na_2O feldspar

Source: H.G. Schurecht et al., "Influence of Fluxes of Spodumene and Feldspar Mixtures on Properties of Chinaware Bodies," *Journal of the American Ceramic Society* 25 (1942): 321–326.

as a partial replacement for feldspar or nepheline syenite in low-temperature (cone 7 or lower) high-flux sanitaryware bodies. The most effective ratio of feldspar: spodumene or nepheline syenite:spodumene is 70:30. Also, the ternary flux, spodumene-talc-nepheline syenite, produces low-fire (cone 4) bodies that have a very high modulus of rupture. The authors applied some sanitaryware glazes on spodumene-containing and normal feldspar bodies using both one-fire and two-fire processes. Commercial plant glazes on the feldspar bodies were as good or better on the spodumene bodies. The authors believed that the small amount of lithium which the glazes absorbed from the spodumene bodies slightly decreased the maturing temperature, thereby producing greater brilliance.

Spodumene is, of course, not the only lithia flux that can be used in whiteware bodies. Lepidolite, which contains both lithia and fluorine, has been widely used. For example, Twells found lepidolite to be an excellent substitute for feldspar in high-tension electrical porcelain and reported that it lowered vitrification temperatures and produced a fine white body.[4] Again, Donahey and Clark report an investigation by R.T. Vanderbilt in which the following cone 8 vitreous artware body is given: clay 65%, dolomite 5%, lepidolite 10%, and Kona feldspar 20%.[5] They also mention a cone 04 vitreous body developed by C.J. Koenig: nepheline syenite 30%, lepidolite 15%, tremolitic talc 5%, flint 5%, and clay 45%.

Lampman replaced the feldspar in whiteware bodies with lepidolite and noticed that fluxing action began sooner.[6] At cone 5, 11% lepidolite was equivalent to 25% feldspar.

Currier, while studying auxiliary fluxes for semivitreous dinnerware bodies, found that combinations of lepidolite and nepheline syenite were superior to feldspar, nepheline syenite, or combinations of these with spodumene.[7] He showed that a combination of 40% lepidolite, 50% nepheline syenite, and 10% talc was the most effective flux for a semivitreous body.

Tumanov and Shvaiko studied the effect of replacing feldspar with beta spodumene, and feldspar and quartz with lepidolite, on the properties of porcelain.[8] They based their work on the porcelain body from the Dulev Factory: 42% Prosyanov kaolin, 4% Oglanlin bentonite, 24% sand, 17% Chupin feldspar, and 13% porcelain body. They showed that the addition of 3.4% spodumene in place of feldspar reduced the firing temperature of the body by 70°C and increased the firing range from 60–70°C to 80–90°C. The addition of up to 6.8% lepidolite in place of feldspar reduced the firing temperature by 30–40° and had no effect on the sintering range. The addition of more than 6.8% lepidolite had no further effect on the firing temperature but it reduced the firing range by 30–40° when compared to the original body. When more than 20% lepidolite was substituted for feldspar and quartz, there was a reduction in firing temperature. A body containing 41% lepidolite was fired at 1150°C to a water absorption of 0.3%. The authors concluded that the addition of spodumene increased the mechanical strength of the porcelain by 60–70%, while the addition of lepidolite increased it by 40–50%. Complete substitution of feldspar and quartz by lepidolite increased the strength by a factor of two.

A recent summary of the use of lepidolite in whiteware bodies was made by Roy and Som.[9] They studied the suitability of Indian lepidolite for developing semivitreous and vitreous bodies. Lepidolite was found to be a more active flux than feldspar; 35% of a combination of feldspar and lepidolite had almost the same fluxing effect as 50% of feldspar. When the flux content was increased from 35% to 45%, the substitution of lepidolite reduced the water absorption by 69% as compared to 57% for the lower flux content. The authors concluded that a combination of lepidolite, feldspar, and talc is highly reactive and well suited for developing vitreous and semivitreous bodies maturing between cone 01 and cone 2.

Lithia as a Low-expansion Filler

The use of free silica for whiteware bodies has been often questioned because of the increased risk of dunting due to the phase transformations of quartz and cristobalite.

Smoke suggested a remedy for this by prereacting flint with alumina and lithia in the proportions: 91-97% SiO_2; 0.75-6.75% Al_2O_3; and 0.25-2.25% Li_2O.[10] The prereacted material was reground and added at the level of 25% to a typical porcelain composition instead of flint. This reduced the volume change that occurs at 573°C due to the quartz transformation.

Another method for reducing the amount of free silica in a whiteware body is to add spodumene and choose firing conditions to promote the following reaction:

$$\underset{\text{spodumene}}{Li_2O.Al_2O_3.4SiO_2} + \underset{\text{silica}}{4SiO_2} \rightarrow \underset{\text{beta spodumene solid solution}}{Li_2O.Al_2O_3.8SiO_2}$$

In this case, the free silica is assimilated in the beta spodumene structure, producing a solid solution that exhibits an even lower thermal expansion than that of beta spodumene. This is similar to the spodumene-clay reaction described in chapter 7.

Commons and Romano added 10% petalite to improve the dunt resistance of an electrical porcelain body composed of 20% ball clays, 30% china clays, 28% feldspar, and 22% flint.[11] They found two compositional areas that produced bodies with a greatly improved resistance to dunting. The first area contained 0–5% Al_2O_3, with flint and alumina varying between 12 and 18%. Bodies in this area matured slightly above cone 8 and were not overfired at cone 13. Firing shrinkages were 9.9% to 10.8%; modulus of rupture reached a maximum of 11,000 to 12,000 psi; and the thermal expansion was 6×10^{-6} (room temperature to 600°C) with no large inflexion at the quartz inversion temperature. The second area contained 5% to 20% pyrophyllite, 2.5% to 15% zircon, and 0 to 15% flint. These bodies matured between cone 8 and cone 10 and were again not overfired at cone 13. Firing shrinkages were 8.5% to 9.6%; modulus of rupture varied from 8800 to 13,500 psi; and the thermal expansion was 5.3 to 6.3×10^{-6} (room temperature to 600°C).

One of the earliest uses of spodumene in porcelain bodies was reported by Bezborodov and Mikhalevich.[12] They developed a composition containing 45.5% Glukhovets kaolin, 4.0% Oglanlin bentonite, 15.0% quartz sand, 24.9% Karelo-Fin bentonite, and 10.6% spodumene. The spodumene contained 5.46% Li_2O. The composition of this porcelain body can be represented by the following molecular formula:

> K_2O 0.219, Na_2O 0.299, Li_2O 0.295, CaO 0.101, MgO 0.086, Al_2O_3 3.775, Fe_2O_3 0.030, SiO_2 16.220, and TiO_2 0.070 or 1 ($R_2O.RO$). 3.805 R_2O_3.16.290 RO_2.

This porcelain body had a thermal expansion of $5.4 \times 10^{-6}/°C$ and a water absorption of 0.05%. Regarding the economics of using spodumene, the authors state that there was no increase in cost because of the reduction of fuel consumption and the decrease in furnace repair costs.

The expansion coefficient reported by Bezborodov and Mikhalevich is lower than that normally associated with porcelain; however, it cannot be regarded as suitable for making heat-shock-resistant ware. Avetikov continued the investigation of spodumene in porcelain bodies but at a level of up to 2% Li_2O.[13] At this lithia content, a body was produced that exhibited a thermal expansion coefficient of 1.5–$2.5 \times 10^{-6}/°C$ from 20°C to 200°C.

Pass and German recognized the importance of a low-thermal expansion for making bodies capable of resisting thermal shock. They added various amounts of petalite (2.4% to 50%) to a standard earthenware body containing 25% ball clay, 25% china clay, 35% flint, and 15% Cornish stone.[14] The thermal expansion was reduced as the amount of petalite was increased. About 50% petalite had to be added to get very low results ($3.0 \times 10^{-6}/°C$ from 20°C to 620°C). The authors believe that, with this type of porous earthenware, additions of 10% to 20% of petalite should have little effect on the expansion due to the alpha-beta quartz inversion. Similar experiments were performed with a vitreous earthenware body of the composition 27.8% ball clay, 26% china clay, 38.2% quartz, 5.5% nepheline syenite, and 2.5% chalk. Here, the addition of 11.2 and 50 weight percent petalite produced expansions (20°C to 620°C) of 8.2 and $4.9 \times 10^{-6}/°C$, respectively. A high-petalite body consisting of 80% petalite, 10% china clay, and 10% ball clay produced a body with an expansion coefficient as low as $2.5 \times 10^{-6}/°C$ (20°C to 620°C).

When considering the above results obtained from petalite-clay mixtures, it should be remembered that petalite cannot theoretically assimilate any more silica into solid solution, whereas beta spodumene can.

Chen patented a hard, dense, mechanically strong, electrically insulating ceramic material based on a microcrystalline mass of synthetic crystals consisting of lithium aluminosilicates and lithium silicates.[15] Although the process claimed is essentially a glass-ceramic one, no nucleating agents are added. The following compositional range was claimed:

Li_2O	4–30%
SiO_2	50–80%
Al_2O_3	3–25%
Fluxing agent	Up to 15%

Motoyuki also recognized the practical applications of a low-expansion porcelain body when he patented a heat-resisting ceramic with an expansion coefficient in the range 3.9×10^{-6} to zero.[16] He states that when petalite is used, it should be added to clay at the level of 40–90 weight percent, but, when spodumene is used, it should be at the level of 30–60 weight percent. Examples of typical results are given in Table 8.2.

The above bodies were coated with low-expansion lithium-containing glazes, and their properties were measured. The bending strength and linear expansion coefficient of these bodies are given in Table 8.3.

Maki and Tashiro were able to extend markedly the firing range of petalite ceramics by adding a powdered lithia glass (Li_2O 10.7, MgO 8.9, Al_2O_3 8.9, SiO_2 71.5 weight percent) to petalite powder in the weight ratio 1:9.[17]

Table 8.2
Examples of Petalite-Clay and Spodumene-Clay Bodies

Body Number	*Petalite*	*Spodumene*	*Plastic Clay*	*Zinc White*	*Silica Sand*
1	44	–	45	6	5
2	65	–	22	3	10
3	70	–	27	3	–
4	85	–	14	1	–
5	–	40	50	5	5
6	–	60	20	–	20
7	50	20	25	–	5

Source: I. Motoyuki, "Method of Manufacturing Low Thermal Expansion Porcelain," U.S. Patent 3,650,817 (March 21, 1972).

Table 8.3
Strength and Expansion of Bodies Shown in Table 8.2

Body Number	*Bending Strength in kg/cm²*	*Thermal Expansion of Body (0–600°C)*
1	650	2.5×10^{-6}
2	600	0.8×10^{-6}
3	600	0.2×10^{-6}
4	550	0.2×10^{-6}
5	650	0.6×10^{-6}
6	550	1.5×10^{-6}
7	600	0.3×10^{-6}

In their slip-casting studies, they showed that the most stable slip composition was 90 petalite, 10 lithia glass, 5 bentonite and 0.02 carboxymethyl cellulose. A modulus of rupture of 900 kg/cm^2 was attained in a slip-cast specimen fired at 1250°C for one hour.

Mueller also investigated the idea of adding a glass to a whiteware composition.[18] He developed a new ceramic material with a low-expansion coefficient for use as a kitchen utensil. A typical body composition is: lithium aluminum silicate glass 30–50%, quartz 15–20%, kaolin 3–4%, and ball clay 5–8%.

Bartlett and Thomas found that very small amounts of lithia were effective in altering the mineralogy of certain compositions in the zirconia-alumina-silica system.[19] A slight change in lithia content had a marked effect on the physical properties of the porcelain into which the zirconia-alumina-silica melts were incorporated. The authors made two electrical furnace melts containing 2.7% and 3.8% Li_2O. The expansion coefficients (20-600°C) were 4.89 $\times 10^{-6}$ for the low lithia body and 2.94 $\times 10^{-6}$ for the high lithia body.

Lithia for Shrinkage Control

As Ware and Russell pointed out, shrinkage that occurs during the firing of vitrified ceramics can create stresses that can induce warpage or cracking, quite apart from limiting dimensional control.[20] Eichbaum and Smoke developed low-shrinkage, vitrified bodies containing spodumene and lead bisilicate.[21] A body composed of 60% spodumene and 40% lead bisilicate exhibited 0.1% firing shrinkage and was impervious after firing to 2020°F. Ware and Russell took advantage of the alpha to beta irreversible phase transformation of spodumene to produce low-shrinkage bodies.[22]

Finally, a dense-sintered ceramic insulating body composed of 1.5–10% MgO, 0.8–3.5% Li_2O, 20–57.5% Al_2O_3, and 40–70% SiO_2 was patented for its low-firing shrinkage (giving an accuracy of ± 0.5% in size).[23]

Raw Materials

The choice of lithium-containing raw material to be used in whiteware production will be dictated by a balance between cost and purity desired. Where there is some tolerance for iron, the less-expensive grades of spodumene will probably be the best lithia sources. These lower grades of spodumene should be particularly useful when lithia is to be used in small amounts as a flux. Where the tolerance for iron is low, either low iron spodumene or lithium carbonate is the preferred lithia source.

Notes

1. J.E. Boyd, Jr., "Pyrometric Properties of Spodumene-Feldspar Mixtures," *Journal of the American Ceramic Society* 21 (1938): 385–388.

2. H.G. Schurecht, J.K. Shapiro, and Z. Zabawsky, "Influence of Fluxes of Spodumene and Feldspar Mixtures on Properties of Chinaware Bodies," *Journal of the American Ceramic Society* 25 (1942): 321–326.

3. C.A. Cowan, G.A. Bole, and R.L. Stone, "Spodumene as a Flux Component in Sanitary Chinaware Bodies," *Journal of the American Ceramic Society* 33 (1950): 193–197.

4. R. Twells, "The Effect of Lepidolite in High Tension Electrical Porcelain Body," *Journal of the American Ceramic Society* 11 (1928): 644.

5. J.W. Donahey and J.D. Clark, "A Neglected Cost-cutting Flux," *Ceramic Industry* (November 1949).

6. C.M. Lampman, "Lepidolite as Flux in Whiteware Bodies," *Ceramic Industry* 33 (1939): 40.

7. A.E. Currier, "Auxiliary Flux Study in Dinnerware Bodies" (Ph.D. diss., Ohio State University, 1948).

8. S.G. Tumanov and V.P. Shvaiko, "Lithium Porcelain," *Steklo i Keramika* 13 (1956): 11–16.

9. S.B. Roy and S.K. Som, "Studies on Indian Lepidolite in Whiteware Bodies," *Transactions of the Indian Ceramic Society* 30 (July–August 1971): 103–110.

10. E.J. Smoke, "Modified Quartz and Method of Making," U.S. Patent 2,726,964 (1955).

11. C.H. Commons, Jr., and P.J. Romano, "Laboratory Development of Dunt Resisting Bodies Containing 10% Petalite," *American Ceramic Society Bulletin* 37 (1958): 353–356.

12. M.A. Bezborodov and P.F. Mikhalevich, "Spodumene Porcelain," *Steklo i Keramika* 11 (1954): 5–9.

13. V.G. Avetikov, "Use of Lithium Compounds in the Ceramic Industry," *Steklo i Keramika* 14 (1957): 10–13.

14. D. Pass and W.L. German, "Whiteware Body Expansion," *Ceramics* (August 1969): 15–19.

15. F.P.H. Chen, "Ceramic Material and Method of Making the Same," U.S. Patent, 3,006,775 (October 31, 1961).

16. I. Motoyuki, "Method of Manufacturing Low Thermal Expansion Porcelain," U.S. Patent 3,650,817 (March 21, 1972).

17. T. Maki and M. Tashiro, "Slip Casting of Petalite Ceramics. Studies of Thermal Shock Resisting Ceramics of the Li_2O-Al_2O_3-SiO_2 System, IV," *Journal of the Ceramic Association of Japan* 72 (1964): 105.

18. C. Mueller, "Lithium Ceramic—A New Possibility for Diversification in the Porcelain and Ceramic Industry," *Sprechsaal* 104 (1971): 920.

19. H.B. Bartlett and R.R. Thomas, Jr., "A Study of the Mineralogical and Physical Characteristics of Two Lithia-Zirconia Bodies," *Journal of the American Ceramic Society* 17 (1934): 17–20.

20. R.K. Ware and R.R. Russell, Jr., "Porcelains Having Low-Firing Shrinkage," *American Ceramic Society Bulletin* 43 (1964): 383–389.

21. B.R. Eichbaum and E.J. Smoke, "Dense Ceramic Bodies of Low Firing Shrinkage," *Ceramic Age* 69 (1957): 22–27.

22. Ware and Russell, Jr., "Porcelains."

23. "Improvements Relating to the Manufacture of Ceramic Insulating Bodies," Britain Patent 1,023,417 (March 23, 1966).

9
Miscellaneous Uses of Lithium

Ferrites

The best-known magnetic mineral is magnetite, or ferrous ferrite. Ferrites may generally be considered as magnetic ceramics having the generic composition:

$$M^{2+} X_2^{3+} O_4$$

Lithium ferrite, $Li_{0.5}Fe_{2.5}O_4$, is an exception. The electrical charge balance is maintained by pairing one Li^+ ion with one Fe^{3+} to produce the equivalent of two divalent ions. It is a unique member of the inverse spinel class of ferrimagnets. Lithium ferrite has a Curie point (the temperature at which a material loses its magnetic properties and becomes paramagnetic) of 637°C according to Hilpert et al.[1] Gorter, however, indicated that measurements on a pure compound gave a value of 680°C.[2] He patented a ferromagnetic material consisting of mixed crystals constituting 17.3 to 43 mole percent of $Li_20.5Fe_2O_3$ and 57 to 82.7 mole percent of $ZnFe_2O_4$.

According to Pointon and Saull,[3] who studied solid state reactions in lithium ferrite, lithium ferrite for square-loop and high-frequency applications has been produced by sintering,[4] hot-pressing,[5] and flash-firing techniques.[6]

Pointon and Saull demonstrated that the sintering of lithium ferrite from lithium carbonate and hematite is promoted by the decomposition of lithium carbonate in the intimate mixture about 400°C below the decomposition point in the pure state. However, above 1000°C there was a loss of lithium and oxygen from the ferrite.

Ridgley et al. also studied lithium oxide volatilization and correlated it with oxygen loss under moderate O_2 pressure.[7] They reported the magnetic moment and lattice parameter of stoichiometric lithium ferrite spinel to be 3736 ± 20 G at 25°C and 8.3296 ± 0.005 Å at 28°C, respectively.

Evaporation of lithium oxide at 1260°C from a lithium-nickel ferrite having the initial molar composition $Li\ Fe_5O_8 . NiFe_2O_4$ was the subject of a brief study by Salmon and Marcus.[8]

Collins and Brown studied lithium ferrites for making latching-type microwave ferrite phase shifters for use in phased-array scanning antennas above C band.[9] As the authors point out, developed materials must not only possess low dielectric losses but must also offer a suitably square hysteresis loop with a low coercive force and a high remanent magnetization. They investigated compounds with the following formulas: $Li_2O.\chi Fe_2O_3 + y\ MnO_2$ (where $\chi = 4.5\text{–}5.0$; $y = 0\text{–}0.6$) and $Li_{(0.5\ +\ \chi/2)}O.Fe_{(2.5\ -\ 3\chi/2)}O_3.Ti_\chi O_2 + y\ MnO_2$ (where $\chi = 0.15, 0.20$; $y = 0\text{–}0.4$). High-density polycrystalline lithium ferrite compositions with suitable dielectric and hysteresis properties were developed.

Weaver and Field examined the possibility of crystallizing lithium ferrite from a glass, thus offering a method for controlling the magnetic and electric properties.[10] The authors concluded that, as the composition of the system $Li_2O\text{-}Fe_2O_3\text{-}SiO_2$ changed from lithium silicate to $LiFe_5O_8$, four different regions were found: (1) stable glasses containing <30 weight percent Fe_2O_3; (2) crystallizable glasses containing about 40 weight percent Fe_2O_3; (3) spontaneously crystallizing melts containing 45–60 weight percent Fe_2O_3; (4) sintered materials containing >60 weight percent Fe_2O_3. Materials in region 1 were paramagnetic while materials in regions 2, 3, and 4 had high dielectric losses, high dielectric constants, and low resistivities. Materials in region 2 had very large coercivities. A limited substitution of Si into the $LiFe_5O_8$ spinel structure was suggested. Weaver patented ferrimagnetic glass-ceramics prepared from lithium ferrite.[11]

Another patent, issued to Research Corporation, claims a mixture consisting of 43–57% SiO_2, 12–18% Na_2O, 6–8% Al_2O_3, 16–31% Fe_2O_3, and 2–6% Li_2O.[12] This mixture was fused and cooled below its crystallization point, resulting in the growth of lithium ferrite crystals.

Grinding Wheels

Houchins found that vitrified bonded, fused alumina grinding wheels with improved properties and performance can be made by including lithia in the alkaline oxide borosilicate bond.[13] According to Houchins, the lithia exhibits a high affinity for the alumina particles so that the latter are wet by the bond at temperatures of 1000°C and lower. In an example, Houchins shows the following composition of the bond:

Flint	750
Feldspar	10
Alumina	6
Boric acid	262
Cryolite	98
Potassium nitrate	45
Fluorspar	3
Talc	28
Lithium carbonate	90

A frit was made from the above composition and ground to 200 mesh. A vitrified grinding wheel was made by molding 91% of 60 grit size fused alumina with 9% of the above frit. After firing at 1000°C for two hours, the vitrified bond of the finished article had the following oxide analysis:

SiO_2	69.10
Al_2O_3	2.91
CaO	0.33
MgO	0.63
K_2O	2.00
Na_2O	3.95
B_2O_3	12.96
Li_2O	3.35
F_2	4.57
Fe_2O_3	0.20

Terada and Yamada increased the grade of grinding wheels by adding 20–25% of petalite to the vitrified bond.[14] Later, Terada investigated the use of spodumene for obtaining vitrified bonds for abrasives made of silicon carbide and fused alumina.[15] He concluded that the best bond for fused alumina was composed of 40% spodumene, 50% feldspar, and 10% "Toseki" fired at 1300°C. For silicon carbide, the best bond contained 30% spodumene, 60% feldspar, and 10% "Toseki" and was fired at 1350°C. The compositions of the three major materials used is shown in Table 9.1.

The investigators who studied the use of spodumene in vitrified grinding wheels took advantage of the previously mentioned eutectic that forms between spodumene and feldspar.

Table 9.1
Composition of Materials Used by Terada

	SiO_2	Al_2O_3	Fe_2O_3	CaO	MgO	Li_2O	K_2O	Na_2O	P_2O_5	LOI
Spodumene	63.6	27.0	0.65	0.21	0.21	7.20	0.25	0.26	0.41	0.46
Feldspar	65.40	19.28	0.06	0.62	0.06	–	9.74	4.66	–	0.21
Toseki	75.98	15.02	0.74	0.14	0.40	–	2.98	0.65	–	3.23

Source: S. Terada and H. Yamada, "Petalite as a Bonding Material for Grinding Wheels," *Reports of Government Industries* (Research Institute, Nagoya) 9 (1960): 341.

Cements

Accelerators are commonly used to increase the initial hardening rate of a cement. Calcium chloride is a widely used accelerator, but it creates difficulties because of its corrosive attack on iron reinforcing rods. Angstadt and Hurley patented the use of an addition of 0.1% to 10% spodumene, based on the dry weight of the cement binder, for accelerating the hardening rate of alite cements.[16] Not only does spodumene provide rapid hardening, but it is noncorrosive. The spodumene can be added to the cement clinker before grinding or it can be added to the cement as a preground powder. To demonstrate the effect of the additive, the authors gave two examples in which a 1000 gram sample of cement was blended with 20 grams of either alpha or beta spodumene and a cement prepared with a water to cement ratio of 0.65 to 1. Compressive strengths, measured after twelve hours, are shown in Table 9.2.

Table 9.2
Compressive Strengths of Spodumene-Accelerated Cements

Concentration of Spodumene	*Compressive Strength, psi*
0 %	181 ± 116
2.0% alpha	638 ± 91
2.0% beta	741 ± 119

Source: R.L. Angstadt and F.R. Hurley, "Spodumene Accelerated Portland Cement," U.S. Patent 3,331,695 (July 18, 1967).

Lithium carbonate is known to accelerate the hardening rate of high alumina cements. However, Men reports that lithium chloride and lithium sulfate retard the setting process of tricalcium aluminate paste.[17]

Mullite Formation

The useful properties of traditional clay-based ceramics are mainly dependent on the development of mullite in the body. The faster the mullite is formed, the sooner the desired properties can be obtained. Parmelee and Rodriguez studied the use of various metallic oxides as catalyzing agents to promote mullite formation in kaolin and found lithia to be a particularly good accelerator at both high and low temperatures.[18]

Shutt pointed out that the formation of mullite at temperatures lower than 1200°C would permit lower firing temperatures and faster firing schedules.[19] He found LiF to be a powerful mineralizer for developing mullite in kaolin at low temperatures. A patent was issued to Fiberglas Canada Limited for the use of 0.5% to 5% (preferably 1-3%) LiF as a mullite accelerator.[20]

Pavlov et al. investigated the effect of 2% of Li_2O, Na_2O, and K_2O on mullite and cristobalite formation in various clays.[21] They concluded that the effect of the additions depends on the mineralogical composition of the clay. Lithium oxide, added as the carbonate, proved to be a strong accelerator.

Finally, Talley studied the effect of lithium carbonate and spodumene on the formation of mullite in kaolin.[22] He reported that the addition of small amounts of lithia can reduce the temperature of the exothermic reactions (which occur in clay at 1031°C and 1216-1261°C) by as much as 119°C and 191°C for the first and second peaks, respectively. Quantitative measurements showed that lithia markedly increased the yield of mullite and reduced the thermal expansion.

High Alumina Refractories

General Refractories patented a refractory high alumina brick mix containing very small amounts of lithium fluoride or lithium carbonate.[23] The resulting refractories possess high cold strength and low reheat expansion. The lithium compound used should be capable of forming lithium oxide on firing. Preferred amounts of the lithium compound are 0.05% to 0.2%.

Densification

Lithium fluoride is a weakened model of magnesium oxide and has been used to densify MgO. Atlas showed that about 0.25% of a lithium salt, particularly LiF, LiCl, or LiBr, lowered the effective sintering temperatures of high purity MgO.[24] He found that the greatest catalytic action of lithium occurred when the magnesias have a particle size of less than 0.5 micron and a surface area of 30 m^2/g or higher. The dense periclase obtained with LiCl-treated magnesia at 1300°C proved to be as resistant to hydration as fused MgO. For this application, lithium has been called a fugitive binder—during firing most of it is lost by volatilization.

Tacvorian proposed a theory that, when a refractory oxide is heated with a compound which is a weakened model, a concentrated solid solution of the compound may be formed in the surface layers of the oxide.[25] This solid solution will have a lower activation energy and will exhibit higher ionic diffusion rates than the pure oxide.

Rice showed that sodium fluoride enhances the densification of MgO as does lithium fluoride. However, temperatures about 100°C higher are required with sodium fluoride to achieve translucency.[26] Lithium fluoride lowers the temperature for hot pressing dense MgO by 300°C to 400°C and imparts some plasticity, according to Rice.[27]

Lithium fluoride has also been used to densify barium titanate.[28]

Battery Uses

During the last few years, a considerable research effort has been expended on the development of high energy density batteries for vehicular applications and military power sources. Because of its high energy density, lithium is a desirable electrode material. However, the successful use of lithium in a high temperature battery is dependent on finding a suitable inorganic separator capable of continuous and protracted operation in contact with molten lithium metal and a fused salt such as the LiCl-KCl eutectic. The separator has two primary functions: to physically separate the positive and negative plates so that electronic current will flow outside the cell and to permit ionic current to flow in the electrolyte between the plates.

Lithium aluminate has been reported to be a suitable separator under certain conditions, particularly at 500°C or below.[29] At higher temperatures (580°C), however, Fishwick and Yeh found a considerable weight loss for lithium aluminate after a 92-hour exposure to molten lithium.[30]

In the sodium-sulfur battery, beta alumina is used as a solid inorganic ionic conductor for sodium ions. Small amounts of lithium are typically used in the manufacture of the beta alumina. For example, Duncan and Hick patented a method for preparing a beta alumina polycrystalline ceramic consisting of Li_2O 0.7-1.5%, Na_2O 8.3–8.9%, MgO 0.5-2.0%, balance Al_2O_3.[31]

Aluminum Pot Lines

Lithium carbonate and lithium fluoride are used in relatively large amounts for increasing the efficiency of aluminum pot lines. The lithium compound, usually the lower cost lithium carbonate, is added to the salt bath where it reacts with the cryolite to form lithium cryolite. The presence of the small lithium cation results in lower power consumption with a higher current efficiency and lower operating temperatures. The latter results in a reduction in fluorine emissions of from 25% to 50%.

Data on the effect of lithium fluoride additions on the efficiency of Soederberg cells may be found in a paper by Wendt.[32] He demonstrated that a lithium fluoride content of 3–4% in the melt of 35–40 K. amp Soederberg pots reduced power consumption and increased current efficiency, thereby increasing production. He also reported on the lower fluorine emissions and speculated on an improvement in lining life. In this respect, it is interesting to compare these data with the data shown in chapter 4 on the effect of lithium in container glass.

Bauer patented a method for making a lithium aluminate with the formula $Li_2O.2\ Al_2O_3$.[33] Such a product is suitable for addition to aluminum cell salt baths.

The traditional method for producing aluminum commercially is the Hall-Heroult process, in which alumina is dissolved in a cryolite bath and reduced electrolytically.

A new process, patented by Russell et al., is based on the electrolysis of aluminum chloride dissolved in a molten solvent with a higher electrodecomposition potential than aluminum chloride.[34] A mixture of equal parts of sodium chloride and lithium chloride was found to be a particularly good electrolyte for this process.

Piezoelectrics

Piezoelectric materials are able to convert mechanical stress into electricity, or vice versa. Single crystals of both lithium tantalate and lithium niobate are useful in piezoelectric and electrooptic devices. Measurements have been made on the crystal structure,[35] optical and electrooptical coefficients,[36] elastic and piezoelectric properties,[37] and thermal expansion coefficients of these materials.[38]

Fraser and Warner stated that the high piezoelectric coupling coefficient and the high Curie point of ferroelectric lithium niobate make it attractive as a transducer material for high-temperature acoustic experiments.[39]

Ouchi et al. patented piezoelectric ceramic compositions having high mechanical strength, high electrical breakdown strength, high electromechanical coupling coefficient and dielectric constant, and high stability in resonant frequency over a wide temperature range.[40] The solid solution range claimed was: PbO 54.8–50.1 mol %, MnO_2 9.3–0.2 mol %, Nb_2O_5 7.3–0.2 mol %, Li_2O 3.8–0.1 mol %, TiO_2 24.8–14.8 mol %, and ZrO_2 26.8–14.8 mol %.

Both lithium niobate and lithium tantalate can be synthesized by calcining lithium carbonate with either niobium or tantalum oxides at the correct stoichiometric ratios. Single crystals are grown by the Czochralski crystal pulling technique.

Paints

Self-curing zinc silicate paints are being used for coating bridges and ships to protect them from corrosion. The paints are composed of a pigment (a fine zinc dust) and a vehicle (an aqueous lithium sodium silicate). According to *Lithium Age,* the aim is to get a high solids content without permitting the zinc dust particles to stick to one another.[41] The paint, when dry, has good water resistance and low water sensitivity. The low and controlled viscosity permits better working qualities even at molar ratios as high as 8:1 (SiO_2:Li_2O), and silica concentrations of up to about 20%.

The preparation of lithium silicate solutions has been discussed in detail in several patents.[42]

Lithium silicate can also be used to coat glasses for the purpose of producing a glare-reducing surface.[43] Here, the lithium silicate has a SiO_2:Li_2O ratio of about 4:1 to 25:1.

Slag for Continuous Casting

In the continuous casting of steel, improved castings can be obtained by providing a protective layer of synthetic slag on the upper surface of the molten steel. An ideal slag should have good chemical and thermal stability and a high solubility for aluminum oxide. Halley et al. have patented a suitable composition that contains 0.5 to 15 weight percent of a lithium compound such as lithium oxide or lithium fluoride.[44] These compositions have a plastic deformation point between 1100°F and 1700°F and possess an alumina solubility in excess of 20% by weight.

Welding Electrodes

Parks patented an electrode composition containing lithium compounds for use in electric arc welding in air.[45] The flux ingredients claimed include 0.2 to 0.5 weight percent lithium by weight of the electrode. The lithium, which is reduced to elemental lithium during the welding operation, can be supplied by the fluoride, silicates, oxalate, aluminate, oxide, chloride, ferrite, or titanate.

Foundry Uses

Soluble silicates are frequently used as bonding agents in cement and refractory compositions. Although aqueous solutions of sodium silicate are commonly the preferred materials, as Aston and Emblem point out, lithium metasilicate has the highest melting point (1201°C) of all known alkali metal silicates and should, therefore, be of considerable interest to foundry technologists.[46]

Efflorescence Control

P.H. Harris found that the addition of 0.1% to 0.5% lithium carbonate to Portland cement eliminated or markedly inhibited efflorescence on masonry.[47] The lithium carbonate proved to be superior to barium carbonate for this purpose.

Catalyst Support

The possibility of using lithium-based catalyst supports for automobile mufflers was mentioned earlier. These are based on crystalline or recrystallized lithium aluminosilicates. Another potential catalyst application, this time for lithium pentaaluminate, was reported by Smith and Beeck.[48] They patented a process for making $Li_2O.5Al_2O_3$ by depositing lithium oxide on the surface of an absorptive alumina.

Fuel Cells

Molten carbonate fuel cells were described in a paper by Tseung and Tantram.[49] Both the insulators and electrodes may have to be made from ceramic materials. These materials, as the authors indicate, must be chemically inert to alkali carbonate melts in the presence of carbon dioxide and of both highly oxidizing conditions (at the cathode) and highly reducing conditions (at the anode) at about 500°C to 700°C. One of the authors' suggestions was to start with gamma alumina and allow it to react to form lithium aluminate. This was then the inert material suitable for making the ceramic components.

Synthesis of Zircon

Ramani et al. reported that Li_2CO_3 accelerated the reaction of SiO_2 and ZrO_2 to form zircon.[50] The lithium carbonate reacted with silica to produce a lithium silicate ($Li_2Si_2O_5$), which then reacted with zirconia to form zircon. The reaction rate was slow below 1000°C, but above this temperature, the lithium silicate melted, making the conversion of unreacted quartz to cristobalite easier; this accelerated the formation of zircon.

Fiber Composites

Levitt recently described a graphite fiber-lithium aluminosilicate composite with a matrix composition of $Li_2O.Al_2O_3.nSiO_2$, where $n = 3, 4$, and 8.[51] The composite has a high volume fraction of unidirectionally aligned graphite fibers and possesses high strength, low density, excellent resistance to thermal shock, and high impact strength.

Notes

1. R. Hilpert, A. Hoffmann, and F. Huch, "Sodium, Lithium, and Copper Ferrites and Their Transformation to Nitrides," *Berichte* 72 (1939): 848–853.

2. E.W. Gorter, "Magnetic Material," U.S. Patent 2,751,353 (June 19, 1956).

3. A.J. Pointon and R.C. Saull, "Solid State Reactions in Lithium Ferrite," *Journal of the American Ceramic Society* 52 (1969): 157–160.

4. G. Blasse, "Crystal Chemistry and Some Magnetic Properties of Mixed Metal Oxides with Spinel Structure" (Thesis, University of Leiden, 1964).

5. R.G. West and A.C. Blankenship, "Magnetic Properties of Dense Lithium Ferrites," *Journal of the American Ceramic Society* 50 (1967): 343–349.

6. Philips Electrical Industries Ltd., "Improvements in or Relating to Methods of Manufacturing Magnetic Cores," Britain Patent 959,643 (January 2, 1963).

7. D.H. Ridgley, H. Lessoff, and J.D. Childress, "Effects of Lithium and Oxygen Losses on Magnetic and Crystallographic Properties of Spinel Lithium Ferrite," *Journal of the American Ceramic Society* 53 (1970): 304–311.

8. O.N. Salmon and L. Marcus, "Note on Sublimation of Lithium from Li-Ni-Ferrite," *Journal of the American Ceramic Society* 43 (1960): 549–550.

9. T. Collins and A.E. Brown, "Low-Loss Lithium Ferrites for Microwave Latching Applications," *Journal of Applied Physics* 42 (1971): 3451–3454.

10. E.A. Weaver and M.B. Field, "Magnetic, Electrical, and Physical Properties of Li_2O-Fe_2O_3-SiO_2 Compositions," *American Ceramic Society Bulletin* 52 (1973): 467–472.

11. E.A. Weaver, "Ferrimagnetic Glass-Ceramics," U.S. Patent 3,694,360 (1973).

12. "Glass Composition Containing Lithium Ferrite Crystals," U.S. Patent 3,492,237 (January 27, 1970).

13. H.R. Houchins, "Abrasive Articles and Method of Making Same," U.S. Patent 2,730,439 (January 10, 1956).

14. S. Terada and H. Yamada, "Petalite as a Bonding Material for Grinding Wheels," *Reports of Government Industries* (Research Institute, Nagoya) 9 (1960): 341.

15. S. Terada, "Bonds Composed of Spodumene, Feldspar, and Toseki For Vitrified Grinding Wheels," *Bulletin of Nagoya Research Institute of Industrial Technology* 12 (1964): 348–358.

16. R.L. Angstadt and R.F. Hurley, "Spodumene Accelerated Portland Cement," U.S. Patent 3,331,695 (July 18, 1967).

17. H.C. Men, "Effect of Electrolytes on Setting and Structural-Mechanical Properties of Tricalcium Aluminate Paste," *Trudy Moskovskogo Khimko-Tekhnologicheskogo Instituta,* no. 68 (1971): 205–207.

18. C.W. Parmelee and A.R. Rodriguez, "Catalytic Mullitization of Kaolinite by Metal Oxide," *Journal of the American Ceramic Society* 31 (1942): 1–10.

19. T.C. Shutt, "Influence of Fluorides on the Formation of Mullite in Kaolin," *Canadian Ceramic Society Journal* 37 (1968): 33–38.

20. "Mullite Formation," Britain Patent 1,238,796 (July 7, 1971).

21. V.F. Pavlov, A.S. Bystrikov, and N.I. Andreeva, "Effect of Additions of Li_2O, Na_2O, and K_2O on the Phase Transformations Occurring on Firing Clays of Different Mineralogical Composition," *Steklo i Keramika,* no. 2 (1970): 38–40.

22. R.W. Talley, "The Effect of Lithia on the Mullitization of Kaolinite" (Senior thesis, University of Delaware, 1968).

23. "High Alumina Refractories," Britain Patent 1,268,334 (March 29, 1972); "Alumina Refractories," Britain Patent 1,268,982 (March 29, 1972).

24. L.M. Atlas, "Effect of Some Lithium Compounds on Sintering of MgO," *Journal of the American Ceramic Society* 40 (1957): 196–199.

25. S. Tacvorian, "Acceleration of Sintering in a Single Phase; Consideration of Mechanism of Action of Minor Additions," *Compte Rendu* 234 (1952): 2363–2365.

26. R.W. Rice, "Hot-Pressing of MgO with NaF," *Journal of the American Ceramic Society* 54 (1971): 205–207.

27. R.W. Rice, "Fabrication of Dense MgO," NRL Report 7334, Naval Research Laboratory, Washington, D.C., November 16, 1971.

28. "A Pressing Way to Better Transducers," *Industrial Research* 32 (December 1972).

29. D.L. Beals and W.H. Mapes, *Separator Materials for the Lithium-Chlorine Battery,* Technical Report Electronic Command 3105 (March 1969); H. Shimotake, A.K. Fischer, and E.J. Cairns, "Secondary Cells with Lithium Anodes and Paste Electrolytes," paper presented at the 4th Intersociety Energy Conversion Engineering Conference, Washington, D.C., September 22–26, 1969.

30. J.H. Fishwick and W.C.T. Yeh, "Ceramic Separators for a High Temperature Lithium Battery," *American Ceramic Society Bulletin* 51 (1972): 633–636.

31. J.H. Duncan and B.K. Hick, "Beta-Alumina Polycrystalline Ceramics," U.S. Patent 3,765,915 (October 16, 1973).

32. G. Wendt, "Operating Experiences with Electrolytes Containing Lithium Fluoride," *Metallurgical Transactions* 2 (1971): 155–161.

33. R. Bauer, "Alumina-Containing Lithium Compounds for Use in Fused-Bath Electrolysis For Producing Aluminum," Britain Patent 1,304,582 (January 24, 1973).

34. A.S. Russell, L.L. Knapp, and E. Hampin, "Production of Aluminum," U.S. Patent 3,725,222 (April 3, 1973).

35. W.L. Bond, "Measurement of the Refractive Indices of Several Crystals," *Journal of Applied Physics* 36 (1965): 1674; A.V. Lepitskii and P. Yu. Simonov, *Zhurnal Fizicheskoi Khimii* 29 (1955): 1201.

36. P.V. Lenzo et al., Appl. Phys. Lett., 8 (1966): 81; G.D. Boyd et al., J. App. Phys. 38 (1967): 1941.

37. R.T. Smith, Proceedings of the 21st Annual Frequency Control Symposium, Atlantic City, April 24–26, 1967; ibid., Appl. Phys. Lett. 11 (1967): 146.

38. Y.S. Kim and R.T. Smith, "Thermal Expansion of Lithium Tantalate and Lithium Niobate Single Crystals," *Journal of Applied Physics* 40 (1969): 4637–4641.

39. D.B. Fraser and A.W. Warner, "Lithium Niobate: A High-Temperature Piezoelectric Transducer Material," *Journal of Applied Physics* 37 (1966): 3853–3854.

40. H. Ouchi, M. Nishida, and K. Nagano, U.S. Patent 3,769,218 (October 30, 1973).

41. *The Lithium Age* (Autumn 1972).

42. R.H. Patton et al., "Protective Coating," U.S. Patents 3,180,746 and 3,180,747 (April 27, 1965); "A Process for the Preparation of Lithium Silicate Solutions," Britain Patent 1,183,120 (March 4, 1970); H.V. Freyhold and V. Wehle, "Method for the Preparation of Lithium Silicate Solutions," U.S. Patent 3,576,597 (April 27, 1971).

43. G.E. Long, F.D. Lititz, and D.W. Bartch, "Lithium Silicate Glare-Reducing Coating and Method of Fabrication on a Glass Surface," U.S. Patent 3,635,751 (January 18, 1972).

44. J.W. Halley, D.E. Grimes, N.T. Mills, and K.R. Yalamanchill, U.S. Patent 3,649,249 (March 14, 1972).

45. J.M. Parks, "Lithium-Containing Welding Electrode," U.S. Patent 3,742,185 (June 26, 1973).

46. J. Aston and H.G. Emblem, "Some Uses of the Soluble Silicates in Foundry Technology," *Metallurgiya* (September 1970): 97–99.

47. P.H. Harris, U.S. Patent 3,188,222 (June 8, 1965).

48. A.E. Smith and O.A. Beeck, "Process for the Production of a Lithium-Aluminum Compound as a Base for a Conversion Catalyst," U.S. Patent 2,474,440 (June 28, 1949).

49. A.C.C. Tseung and A.D.S. Tantram, "The Use of Ceramics in Molten Carbonate Fuel Cells," *Science of Ceramics* 4 (1968): 367–379.

50. S.V. Ramani, S.K. Mohapatra, K.V.G.K. Gokhale, and E.C. Subbarao, "Zircon-A Review (Part 1)," *Transactions of the Indian Ceramic Society* 30 (1971): 9–32.

51. S.R. Levitt, "High-Strength Graphite Fiber Lithium Aluminosilicate Composites," *Journal of Material Science* 8 (1973): 793–806.

10
Analysis of Lithium

Most investigators who are studying the effect of lithium or spodumene in ceramic systems will at some time need to determine the amount present in the final ceramic or glass. In general, determinations will fall into one of two categories—refractory users may need to know the amount of alpha or beta spodumene in their bodies, and other users may want to quantitatively determinate the total lithia.

Alpha-Beta Spodumene by X-ray

The percentage of alpha or beta spodumene in a mineral concentrate or body can be determined by the disappearing x-ray line technique. Here, standards are made containing various amounts of the two minerals, and these are subjected to x-ray diffraction analysis for the two "d" lines at 4.21 and 2.93. These lines were found to be the most sensitive to changes in mineralogy. The method is rapid (although samples should be finely ground for incorporation into the sample holder) and accurate to within a few percent of beta spodumene.

Wet Analysis of Lithia

Manufacturers of glasses, glass-ceramics, glazes, enamels, and whitewares will find flame photometric analysis to be the best procedure for determining lithia in their bodies. First, the sample is dissolved; then the solution is burned, and lithium is determined by flame photometry or atomic absorption at a wavelength of 6708 Å.

Dissolution of the sample is normally performed by an acid digest using a mixture of perchloric or sulfuric acids and hydrofluoric acid. The samples should be finely ground to aid solution, and any alpha spodumene present in the sample should be converted to beta spodumene by calcination at 1800°F for a half-hour. During the digest step, the solution should be taken down to strong fumes to remove flurosilicic acid and excess HF. For difficulty-soluble samples, such as those high in alumina, the digest should be performed with sulfuric and hydrofluoric acids, and any residue should be fused with potassium pyrosulfate and leached again. Samples containing alkaline earth oxides (those which give insoluble sulfates) should be dissolved in a mixture of perchloric and hydrofluoric acids.

The method is accurate to ± 1.5 Li_2O of the amount present.

Index